CHILDREN´S WAYS OF HANDLING AND EXPERIENCING NUMBERS

GÖTEBORG STUDIES
IN EDUCATIONAL SCIENCES 113

Ann Ahlberg

CHILDREN´S WAYS OF HANDLING AND EXPERIENCING NUMBERS

ACTA UNIVERSITATIS GOTHOBURGENSIS

ISBN 91-7346-315-9
ISSN 0436-1121

Distribution: ACTA UNIVERSITATIS GOTHOBURGENSIS
BOX 5096
S-402 22 Göteborg, Sweden

Abstract

Title:	CHILDREN'S WAYS OF HANDLING AND EXPERIENCING NUMBERS
Language:	English
Keywords:	Number concept, numerosity, ways of handling numbers, experience of numbers, the dynamics of learning numbers, phenomenography, awareness
ISBN	91-7346-315-9

This report deals with the different ways in which children handle and experience numbers and outlines a study carried out within the phenomenographic research framework. The main focus is on describing and analysing the different ways in which children handle numbers and, with this as a base, showing how they experience the meaning of numbers in a problem solving situation. 38 six year old children, were interviewed and videotaped solving different types of mathematical word problems.

The results show that the children handle numbers in many different ways, but that these can be grouped into five main categories; Saying numbers - Estimating - Counting - Structuring - Using known facts; and although the way children handle numbers simultaneously constitutes their experience, it is possible to distinguish between ways of handling numbers and ways of experiencing numbers on an analytical level. The ways of handling numbers identified thus served as a basis for an analysis and interpretation of children's different ways of experiencing numbers and it was found that children experience numbers in four different ways, namely as Number Words - Extents - Positions in Sequence - Composite Units. The different ways in which children handle numbers were found to influence their interpretation of the meaning of number. However, this was not in a one to one correspondence. Some ways of handling numbers were found to be related to more than one meaning. Consequently, children can develop the same understanding of numbers by different ways of acting and thinking.

One conclusion drawn from the study is that understanding numbers and learning arithmetic skills is not only a question of the quantification of objects. Neither is it merely a matter of learning how to count on the number sequence or of developing logical thinking. It is, rather, a question of exploring the different aspects and possible qualities of the numbers - of experiencing numbers. In other words, experiencing in the sense of simultaneously perceiving different aspects of numbers.

There are many implications for the teaching process. In order to better develop their understanding of numbers children should group them in various ways by using sensuous perception and by touching, hearing, and seeing simultaneously, experience the numbers in a wholeness consisting of smaller parts. These perceptual experiences should be integrated in a social context in which children talk and reflect upon the different ways of experiencing various aspects of numbers. By using various ways of handling numbers in different contexts, they can be aware of different aspects of numbers and develop their understanding of their meaning.

Contents

Chapter 1

DIFFERENT VIEWS OF THE DEVELOPMENT OF NUMBER CONCEPTS 1

Subitizing - the Immediate Perception of Small Numbers 3
Logical Thinking and Reflective Abstraction 4
Counting Principles and Innate Number Knowledge 7
The Sequence of Whole Numbers 9
Finger Counting - Sequential and Analogical Representations 11
Problem Solving Strategies 12
Learning Numbers in an Experiential Perspective 14
Exploring Different Aspects of Children and Numbers 21

Chapter 2

AIM AND METHODOLOGY OF THE STUDY 23

Method and Theoretical Framework 24
The Phenomenographic Interview 26
Questions about Validity and Reliability 27
The Empirical Study 29
Interviews 29
Data Analyses 30
Tasks 31

Chapter 3

RESULTS 34

Children's Ways of Handling and Experiencing Numbers 34
Ways of Handling Numbers; Saying Numbers 35
Experiencing Numbers as Number Words 42
Ways of Handling Numbers; Estimating 48
Experiencing Numbers as Extents 50
Ways of Handling Numbers; Counting 51
Experiencing Numbers as Positions in Sequence and as Composite Units 64

Ways of Handling Numbers; Structuring 70
Experiencing Numbers as Composite Units 73
Ways of Handling Numbers; Using Known Facts 74
Experiencing numbers as Number Words and Composite Units 75
Children's Ways of Handling Numbers Revisited 77
Children's Main Ways of Handling Numbers 80
Handling Numbers in More Than One Way 81
Children's Ways of Experiencing Numbers Revisited 83
The Nature of the Variation 84
The Sensuous and Simultaneous Experience of Numbers 87
Experiencing Numerosity by Keeping Track and Marking 88
One Way of Handling afford more than one Meaning 89

Chapter 4

THE DYNAMICS OF EXPERIENCING NUMBERS 91
Looking at Number Development through the Experience of the Individual Child 91
From Number Words to Composite Units 93

Chapter 5

CHILDREN LEARNING NUMEROSITY 97
The Results in relation to Different Views on Learning Numerosity 98
Learning Numerosity in Educational Settings 104
Bringing about Changes in Children's Ways of Experiencing Numbers 105
Bringing Children's Experienced World into the Teaching Process 105
Concluding Remarks 108

REFERENCES 110

APPENDIX

Acknowledgements

This is the third publication within the research project "Numerosity and the development of arithmetic skills among visual impaired children, hearing impaired children and children without these impairments". Earlier publications are "Grasping numerosity among blind children" (Ahlberg & Csocsán, 1994) "Conceptions of Numbers -The perspectives of Hearing Impaired Norwegian Schoolchildren" (Frostad & Ahlberg, 1996). In preparation is "Blind children's experience of numbers" (Ahlberg & Csocsán). A fifth summarising report including a comparison between the three different groups of children is in progress.

The research was financed by The Swedish Agency for Education and would not have been possible to conduct without the co-operation of the teachers and above all the children. I am mostly thankful to them for trying to answer all questions. Also, my gratitude goes to Ference Marton who has shown a great interest in the work and I want to thank him for offering helpful comments on an earlier draft. Per Frostad, who is participating in the research concerning hearing impaired children, also has influenced my thinking and given valuable comments on an earlier draft. Other people who has discussed the work is Emmy Csocsán, participating in the research concerning blind children and my colleagues Elisabet Doverborg, Jan-Åke Klasson, Bengt Persson and Ulla Runesson. I am grateful to Anna Klerfelt who has interviewed the children and Dennis Beach for making my English readable.

Mölndal, June 1997

Ann Ahlberg

Chapter 1

DIFFERENT VIEWS OF THE DEVELOPMENT OF NUMBER CONCEPTS

How do children come to their understanding of numbers? This is the question focused upon in this report. In introducing this question an overview is given illuminating some parts of the extensive body of research which has been carried out concerning how children understand numbers and learn arithmetic skills. Learning to count and to develop arithmetic skills is one of the major field of research in the area of child development. Fuson (1992) states that research has yielded a description of a developmental progression of increasingly abstract and efficient conceptual structures that children construct in order to perform addition and subtraction. However, there is not as much known as to how and why children move through these stages or processions or how teachers can best help them to do so.

This report aims to contribute to answering the question of which pathways children use in order to come to an understanding of numbers. It will attempt to add to the existing knowledge of the growth of mathematical thinking through a process oriented description of children's early mathematical understanding and to formulate guidelines with which to aid both teaching and children's learning of mathematics in the classroom.

Children's understanding of numbers is investigated in a problem solving context, an approach frequently taken in earlier studies. The main difference however, between this study and earlier ones is that in this investigation the analyses of the children's different problem solving strategies - or in the terminology used in this study, "ways of handling numbers" - is only taken as the point of departure for the analyses and serves as the base for an interpretation of how children experience the meaning of numbers.

Although children's ways of handling numbers does, at the same time, constitute their actual experience, revealing a distinct difference between the description of how children go about solving problems and the various ways in which they experience and understand the meaning of numbers is analytically possible.

A huge amount of research has been carried out within different research traditions in which various views are found both regarding what constitutes a conceptual understanding of numbers and how this understanding is developed. What kind of theories, methods, or language do the various views turn to in the attempt to explore children's understanding of numbers? I will try to point out some of the similarities and differences between a few of these views and to discuss in which ways they attempt to add to the valid body of existing knowledge concerning children and their understanding of numbers.

Within the field of mathematics education there is an increasing awareness of the significance of epistemological and philosophical issues with regards to important traditional areas of inquiry (Ernest, 1991, 1994; Mellin-Olsen, 1987). The intention of the following report however, is not to outline these areas of philosophy of mathematics education or to polarise between different theoretical standpoints (such as information processing systems, constructivism, or situated cognition), but rather to describe some of the themes and aspects regarded to be the most important features in describing children's development of numerosity within different theoretical standpoints.

In the body of research concerning the learning and teaching of mathematics the perspectives vary from models trying to unify knowledge of skills and procedures with conceptual knowledge (Gray & Tall, 1994; Hiebert & Lefevre, 1986) to perspectives which describe the importance of socio-cultural differences between "the culture of school mathematics" and the use of mathematics in everyday life (Brown et al, 1989; Carraher et al 1985; Lave, 1988, 1993; Resnick 1987, 1989; Säljö & Wyndham, 1987). Furthermore, problem solving is often emphasised as an important tool in the learning of mathematics, especially regarding older students; (Ahlberg, 1992; Lester, 1988; Schoenfelt, 1987;) and the use of spoken, written, and pictorial modes of language in the learning of mathematics is emphasised (Ahlberg, 1992; Johnsen Høines, 1990; Pimm, 1987).

In research directed towards the task of describing the different ways in which children develop understanding of numbers the perspectives vary from a genetic point of view, looking at numerosity as a biologically anchored ability (Gelman & Gallistel, 1983, 1992; Wynn, 1992), or the notion that numbers may arise from the activity of counting itself (von Glasersfeld, 1993,1994; Steffe et al, 1983), to stressing that counting on the number sequence is the most important tool in the development of number sense (Fuson 1992). In some models the children's use of fingers to represent numbers is heavily emphasised and looked upon as a tool with which to develop numerosity (Brissaud, 1992; Marton, 1988; Marton & Neuman, 1989; Marton & Booth, 1997; Neuman, 1987, 1989). Despite the fact that most researchers agree that the development of numerosity is a very complex process, different views, basing their focus upon different specific aspects, can be identified in the huge number of approaches outlining how children come to understand the meaning of numbers and develop numerical knowledge.

Subitizing and the Immediate Perception of Small Numbers

The ability to perceive and differentiate sets of one, two, and three objects immediately - without counting the units one by one - is called subitizing and is argued to be a biologically anchored perception of patterns, patterns that can be seen or heard as representations of numbers (Antell & Keating, 1983; Klahr & Wallace, 1973). Fischer (1992) argues that comprehending patterns can play an important role in concept development, as does Neuman (1987) who claims that the ability to subitize is an important component in the development of number concepts and that by making the numbers perceptible on their fingers children can expand the subitizing range and that this, in turn, has great importance towards their comprehension of numbers.

In contrast to this position the radical constructivist von Glasersfeld (1993) claims that, conceptually, it is one thing to recognise a spatial or temporal pattern as a pattern one knows and has associated with a certain name, and quite another to interpret the pattern as a collection of unitary items that constitute a certain numerosity. The connection between patterns and numbers is of a semantic nature and it is not until a link is constructed between different figural patterns and the first number words that a child has a tool with which to develop the embryonic ability to subitize into a fully grown sense of numbers. For every number word however, there can be a great variation of patterns with a semantic connection to each number, and because of being linked to a figural pattern, a counting word is not automatically turned into a numerical concept. This demands "reflective abstraction" to accomplish and a child's ability to subitize is thus, according to von Glasersfeld, not directly connected to the development of numerosity. On the other hand, perceptual configurations provide a necessary precondition for this abstraction process.

Logical Thinking and Reflective Abstraction

"Reflective abstraction" - is a notion Piaget (1969) originated to denote the construction of knowledge through reflection upon objects and actions, successively increasing the ability for abstract thought. Piaget is a key figure within research concerning number concepts and a large amount of the research on children and their development of mathematical skills has been dedicated to verifying, reinterpreting, or discounting his theories. A great deal of the studies concerning the understanding of mathematics emerges from Piaget´s work and tackles the questions of classification and conservation.

The sequence of numbers is indissociably cardinal and ordinal in Piaget's view and all mathematical understanding has its roots in the growth of logical ability and develops in accordance with other conceptual knowledge. Cardinality or the manyness of a set builds on class logic. All sets, where the elements are in one-to-one correspondence with each other, will have the same numerosity. Classes are included in classes leading to the additive composition of numbers. The class four is included in five, which in turn is included in the class six. Cardinal numbers are in themselves conceptual units that can be decomposed into compositions of smaller cardinal numbers. Ordinality, on the other hand, builds on the logic of asymmetrical relations, where it is the systematic seriation of cardinal numbers which in an increasing or decreasing order thus make up the ordinal aspect of numbers. There is a contradiction between class logic depending on similarities between the elements in a class and asymmetrical relations depending on differences between the elements in a series. However, when this system is applied on sets, a fusion between class inclusion and seriation takes place and they form a united, operational totality, where the different classes are only separated from one another according to the number of elements they include.

Piaget did not devote much attention to early numerical activities such as counting and claims that a child does not have a developed understanding of numbers until a fusion of the notion of cardinality and ordinality takes place. In accordance to this he has argued that children cannot learn the compositional relationships between different numbers or engage in numerical reasoning until about seven years of age. Several researchers have also shown that the relationship between children's ability to count and their understanding of conservation is not very strong and that long before they are able to conserve numbers children can exhibit good counting ability (Gelman & Gallistel 1983; Wynn, 1992).

von Glasersfeld (1983; 1993) and his colleagues (Steffe & Cobb, 1988; Steffe, Richards & Cobb, 1983) have outlined a model of the development of children's early mathematics within a constructivist theory. They suggest that all quantification and numerical thinking has its roots in counting discrete units which can be repeated and combined. In order to have any meaning the number sequence must be created by reflective abstraction, that is by being isolated from the units or items the children have originally or primarily identified. The model is built upon the construction of counted units and the way in which these units appear in the mind of the counting child during different developmental stages. Five types of countable items are described:

1. *Perceptual unit items* are counted when the children need a collection of objects in order to carry out the counting activity.

2. *Figural unit items* are visualised images of perceptual items.

3. *Motor unit items* are constructed when children count independently of the immediate perceptual surrounding and, for example, point over a hidden number of items.

4. Verbal *unit items* are created as the word is substituted for the preceding types of countable items.

5. *Abstract unit items* are constructed when the child realises that any of the previous types of unit items may be taken as units to be counted. When knowing that a number word has a specific position in the number sequence, as well as being the number of collection of discrete unitary items that could be co-ordinated with the numbers, the child has an abstract conception of numbers.

In contrast to Piaget, "counting" is, in this model, accorded an important role in the formation of number concepts, as the activity of counting on the number sequence provides the material for reflective abstraction. According to von Glasersfeld (1993) the word "counting" is used in a serious, practical context and requires an explicit indication of the numerosity of a given set of items. In order to satisfy these conditions, determining a specific numerosity demands the ability to:

- distinguish discrete *countable* items,
- recognise the *plurality* of such items as the collection to be counted,
- have an ordered, sequence of individually different "tags", for example the standard number word sequence,
- co-ordinate the elements of this sequence in a *one-to-one* fashion with the items of the collection,
- and to be aware of the *symbolic function* of the number words used - at least to the extent that apart from serving as tags in the counting, the last number word used can be interpreted as an indicator of the counted collection's numerosity. (This last condition has been called "a sense of number" by some and "cardinality" by others.) (von Glasersfeld, 1993, p. 230).

All quantification and numerical thinking is thus based on the construction of discrete units which can be repeated and combined. The child has to isolate discrete experiential units, co-ordinate these to the numbers, and, through reflective abstraction, give meaning to the whole number sequence.

Counting Principles and Innate Number Knowledge

Gelman and Gallistel (1986) suggest that the ability to understand numbers is originally present in the minds of human beings and that children's number sense builds on counting, that early counting is principled and that children's counting is guided by mental representations of these principles defining correct counting. Children count on the number sequence and as they grow older, successively learn the meaning of the number symbols.

To understand numbers children must understand five fundamental principles guiding counting:

1. The *one-one principle* states that each and every item in a set must be marked with one, and only one, unique tag.
2. The *stable-order principle* specifies that, whatever the tags, they have to be used in a repeatable, ordered way.
3. The *cardinal principle* stipulates that the last tag marking the last item in a set represents the total number of items in the set.
4. The *abstraction principle* specifies that various kinds of items may be collected together for the purpose of counting.
5. The *order-irrelevance principle* states that the items in a set can be marked in any order and, as long as the one-one principle is observed, the outcome will be the same. (Gelman & Gallistel, 1986, p. 77-82).

Understanding of the five counting principles develops continuously as the child grows older; however, the primary comprehension must precede skilled counting, for even if the children use number words in different contexts and count using the number sequence, it is not until they have understood and can apply all five principles that they can be said to have developed a deep understanding of numbers and counting. Coherent with this,

Gelman and Meck (1992) argue that early numerical competence is better explained by positing a counting process than by relying on pattern recognition or development through a subitizing process.

The thought that children's learning of mathematics should be underpinned and directed by counting principles has been criticised by many researchers, Baroody for instance (1992), who concludes that existing research does not settle the debate concerning the developmental relationship between counting principles and counting skills. Baroody indicates however, that available evidence does suggest that children do typically understand the principles adduced by Gelman and Gallistel (1986) some time before they begin school, but it does not show whether such understanding exists prior to the development of any counting skill and it is not clear whether children have an innate non-verbal counting mechanism rather than an innate subitizing mechanism, that innate counting mechanism require all five counting principles, or that such principles actually guide the initial development of number word counting skills.

Wynn, (1992 b), has shown that children as young as eight month of age are capable of engaging in numerical reasoning, although they have not learnt any counting words. Two year old children know that the counting words each refer to a distinct numerosity, though they do not yet know to *which* numerosity each word refers. She suggested an alternative to the model of innate number knowledge proposed by Gelman and Gallistel (1986) based on the idea that the numerosity of a set of events or objects may be non verbally tracked and appreciated by a "counting mechanism" such as an accumulator, which is filled up in equal increments for each entity "counted", resulting in a magnitude representation. In this system, numerosity is embodied in the structure of representative states of the accumulator which are in one-to-one correspondences with the entities counted and arrived at in the same order, as each increment is a cardinal representation of the entities counted so far. Cardinality is, in this sense, inherently embodied in the structure of the representation and the output value of the accumulator resembles the physical configuration of the sets of elements to be "counted". Wynn (1992 b, p. 226) suggests that there are two essential components to understanding the cardinal meaning of a particular number word:

1. The knowledge that a number word refers to a *numerosity*, regardless of which it refers to.

2. The knowledge of the precise numerosity a word refers to (its cardinal meaning)

Wynn´s model suggests that in order to interpret the linguistic counting words children must map the information between two very different kinds of representations - their own magnitudinal representations and the sequential representation of the standard number string - which is quite an undertaking for small children.

Gallistel and Gelman (1992) propose that the preverbal counting mechanism outlined by Wynn is the source of the implicit counting principles guiding the acquisition of verbal counting which they outline in their own model. However, Wynn declares:

> ... the finding that children know that the larger number words refer to numerosities before knowing which numerosities, they refer to, even for words that are well within their counting range, conflicts with the prediction of the "counting principles" theory that once children realise that the number words refer to numerosities, they will link them up with their own, already ordered list of mental number tags and hence will know the precise numerosity (Wynn, 1992 b, p. 245).

The Sequence of Whole Numbers

Fuson (1992) emphasised the importance of counting on a number sequence to develop cardinality and that young children have to represent numerical situations in a concrete manner using objects, before they can become able to represent numerical situations on the number sequence by using the number words as objects. There are a lot of models suggesting that counting on the number sequence contribute to children's development of numerosity. A strategy sometimes used as a purely mental act is "double counting" (Fuson & Hall, 1983; Steffe et al, 1983). This means, for example, that in order to come up with an answer to the task

2+_=9 the child counts the number words in two parallel number sequences like this: 3 is 1; 4 is 2; 5 is 3; 6 is 4; 7 is 5; 8 is 6; 9 is 7.

Fuson (1992) outlined a model for the development of the knowledge of the number sequence consisting of four levels. At the first level, the "Unbreakable Chain level", initially the number words are not separated from each other and children understand the number sequence as a string with an unbreakable list of names. Gradually, as the number words are differentiated and the words can be paired with objects, counting obtains its cardinal result.

At the second level the "Breakable Chain level", children are able to start counting at any point in the sequence, but still need perceptible units to be counted. At this level the children are able to see a set as composed of subsets, and are able to let the counting of the elements in the subsets be embedded in the sum count. At the third level, the "Numerical Chain level" the sequence words themselves become the objects for counting and children can start the count from the number of elements in the first subset and count the number words corresponding to the number of elements in the second subset.

Finally, at the fourth level, the "Bidirectional Chain level", the sequence becomes a unitised, seriated, embedded, bi-directional, cardinalized sequence where the two addends can be differentiated from each other as well as from the sum. Relations between the two different addends can be elaborated, structures can be established and the addends can be divided. On this level the children can construct relationships between two different addend-addend-sum situations, chunk addends into parts for more convenient adding or subtracting, decompose the cardinal number into ideal unit items, compare and relate cardinal numbers and have cardinal numbers embedded within the number word sequence so that they are seriated and included within each other.

Thus the number word sequence moves from being a rote, meaningless series of utterances to being constituted by number words that come to stand for objects in the counting procedure and that can take on cardinal meaning as the sum of the words earlier in the sequence. According to Fuson it is not until reaching this last level of understanding of the number sequence the children have a genuine understanding of cardinal numbers.

Finger Counting - Sequential and Analogical Representations

Several studies show that using fingers to grasp the numerosity of numbers in making sequential or analogical representations of numbers is fairly common (Ahlberg, 1992; Ahlberg & Hamberger, 1995; Brissaud, 1992; Fuson, 1992; Marton, 1988; Marton & Booth, 1997; Neuman, 1987; Steffe & Cobb, 1988).

In a model of developmental levels of addition and subtraction solution procedures Fuson (1992) show the ways in which children use their fingers solving canonical and non canonical problems. Children are counting out their fingers as objects and there is a development from "counting all" to different ways of "counting on" and "counting down". For most children counting down is considerably more difficult than counting forwards and the sequence - counting - cardinal relationships are also somewhat more difficult to establish in the backwards procedures. Fuson also points out the effect of cultural influences on the children's use of fingers doing arithmetic.

> The structure of the number - word sequence and the way in which fingers present addition and subtraction both seem to affect the developmental sequence of solution procedures used by children in a given culture (Fuson, 1992, p. 147).

Brissaud (1992) claims that children construct numerical knowledge gradually by forming analogical sets through one-to one correspondence. When a child represents a set of four items by creating a set of four fingers through one to one correspondence, manyness is represented by another manyness and enables the child to recognise the four items in the set both as individual items (or individuals) and as forming a whole (unification of individuals). In verbal counting on the other hand, the numerosity of the sets is represented in a much more indirect way.

Neuman (1987) and Marton & Booth (1997) propose that a great many seven year old children use their fingers to make patterns and to create "finger numbers" when solving verbal problems and that this procedure makes it possible for them to perceive the numbers without counting and represent wholes and parts within the whole. For example, in solving the problem 2+_=9 the child starts by raising 9 fingers palm up, opening both hands and folding in a thumb. Looking at the fingers, the child then folds in two more - the two next to the thumb - and sees that there are now 7 fingers raised. According to Marton and Booth (1997, p. 65) children can simultaneously experience the ordinal and cardinal aspects of numbers when using their fingers in this way.

Problem Solving Strategies

In the research described above, most researchers agree on the importance of counting on a number sequence for the children's development of number concepts. Furthermore, that children use the number sequence counting upwards and downwards when solving simple addition and subtraction problems. However, before counting on the number sequence using the number words as objects children need physical objects to model verbal problems. Carpenter and Moser (1982) have made an analysis of the strategies used in solving different types of elementary arithmetic problems and claim that children primarily solve simple addition and subtraction problems by modelling with physical objects or with their fingers. Children use different objects to represent the numbers and manipulate them in a one-to-one correspondence with the mathematical content of a task. This kind of correspondence modelling is used for addition as well as subtraction. For instance, when a child works out 3+5 they initially count the first three objects, then they count five objects and finally they count all the eight objects in the set. This addition strategy is called "counting-all-with-models".

On the next level of knowledge development, children leave the direct correspondence with physical objects and instead use different strategies for counting on the number line. One way to solve 3+5 might be "counting on from the first", and then counting starts at four, followed by five, six, seven and eight. The most advanced strategy in addition is "counting on from larger". The child then starts counting from the largest number, i.e. starts at six followed by seven and eight.

Children also use their knowledge of number facts. Sometimes they use a "known fact" involving the numbers given in the problem. In other problem situations some children use "derived facts" generated from a small set of known basic facts. For example, if a child knows that 3+3 is 6 and realises that to solve 3+5 there are 2 more to be added, then according to Carpenter and Moser they are using their knowledge of derived number facts.

Siegler & Jenkins (1989) describe differences in individual preferences of strategy choice through micro-genetic studies where they try to explore the discovery and the generalisation of strategy use. They emphasise that the process of strategy construction takes place over time. Sometimes the process takes only a few seconds or minutes, but for many of the most interesting strategies, it takes weeks, months or even years. Considering the process of strategy construction along a dimension of time suggests a basic division of the process into two periods; an initial period of strategy discovery and a later period of strategy generalisation. People often learn new strategies even when existing ones produce consistently correct performances. The use of a strategy in a domain is an important milestone, but by no means the end of the strategy construction process. A strategy is not fully constituted until a person extends it to the entire range of situations in which it can be useful.

Very small children have an implicit knowledge of counting and arithmetical principals long before they verbally can formulate their knowledge (Gelman & Baillargeon, 1983; Gelman & Gallistel, 1986; Wynn 1992) and develop their mathematical thinking in their daily life and have an ability to solve mathematical problems before formal teaching begins. Ginsburg, (1983) puts forward that five and six year old children often use purposeful informal strategies when solving problems in every day life, but their methods are often intuitive and differ from the mathematics they meet in school. As described, children use objects modelling the problems in different ways but they also count in their heads or on the fingers.

The problem solving which children perform in every day life without being aware of it as "mathematics" is considerably important when it comes to understanding the meaning of numbers. Carpenter and Moser (1982) emphasise that the meeting with formal school mathematics, with its well defined methods and procedures, is a very critical stage in learning of mathematics, where children have to abandon their own ways of thinking in favour of the methods used in school. According to Baroody (1987) this shift between informal and formal methods can influence their understanding and attitude towards mathematics in a negative direction and may have bad consequences for their future learning.

Learning Numbers in an Experiential Perspective

Problem solving in every day life is of considerable importance for the growth of mathematical understanding as a child's learning of numbers and counting is not only linked to formal teaching. Before attending school most children verbally express different understandings of numbers and counting (Ahlberg, 1995; Gelman & Gallistel, 1983; Hughes, 1986; Ginsburg, 1977; Wynn, 1992).

Ahlberg and Hamberger (1995) interviewed 103 six year old preschoolers, who had not attended any formal instruction in mathematics concerning their thoughts on the meaning and function of numbers and counting. They showed a wide variation of ideas and found that these children understood the meaning of numbers and counting in six main qualitatively different ways:

Saying a string of words

The children refer to string of words without any connection to the number words or they say words with some links to numbers.

A few children refer to words not connected to the numbers mentioned the alphabet saying:

"A, b, c, d, n, m, n, o ...".

Others say a nursery rhyme including numbers. The children give a lot of different examples like[1]:

> "One, two, buckle my shoe, three, four, shut the door".

Using numerals

In every day life we most often do not make a sharp distinction between numbers, numerals, number words and figures. The children use these terms mainly with the same meaning. Some of them experience that counting means using numerals.

You can *read* numerals:

> "There are numbers in books, so that one should know where one are".

You can *write* numerals in books or on a computer:

> "Counting is to write numerals as in books".

Using number words

Some children experience counting as saying number words, not in the order of the standard number sequence, but experienced as separate number words.

A few children simply refer to some numbers:

> "You can tell 8, 5, 2, 11".

It is rather usual that the children connect counting with age simply saying:

> "You can count how many years you are".

Some children give examples as:

> "30 years, 90 years, 28 years, 20 years, 30 years, 40 years".

Counting on the number sequence

The meaning of counting is most often experienced as counting on the number sequence. The children refer to the number sequence in a lot of different ways.

Counting forwards:

> "If you start at 1, you continue 2, 3, 4, 5, 6".

Counting backwards:

---

1 This is an English rhyme compatible to the Swedish rhymes the children said. For example "Ett, två, tre, fyra, alla byxor äro dyra" or "Katten, musen, tiotusen."

"9, 8, 7, 6, 5, 4, 3, 2, 1.

Counting on the ordinal numbers:

"The first, the second, the third".

Counting equal numbers:

"2, 4, 6, 8, 10".

Counting on the tenths:

"10, 20, 30, 40, 50".

Counting in English[2]:

"One, two, three, four, five, six ..."

Performing calculations

Some children refer to numbers as doing addition and subtraction and sometimes simply say that you add or take away. However, most common is to refer to operations within a smaller number range and give various examples.

Some children just say that you add or subtract:

"Yes, plus and things like that, I do not know how but I know that you can do it".

A few children give a lot of different examples:

"You calculate 1+1=2 or 5+5=10 or 5-2=3".

It is usual that the children refer to "doubles" such as "2+2=4", "5+5=10". Sometimes "a double" can be in a bigger number range such as "20+20=40".

Counting sets of elements

The children experience counting as counting quantities. They refer to people or objects, a single child mentions that you can count how many fingers you have. Usually the children mentions that you can count the number of children in the group in pre- school, marbles or cars.

"I can count how many teeth I have lost".

"You can count how many papers there are ... how many shoes you have".

---

2 English was the only foreign language the children used.

In summarising the results it could be said that all children in this study experience two or more ways of understanding numbers and counting. Most common is that they refer to "counting on the number sequence" - 86 % of the 103 interviewed children gave this as the first answer. In spite of not expressing their understanding in mathematical terms, it is obvious that six years old experience different aspects of numbers ought to be used in various situations in their daily lives.

Also, in the same study, Ahlberg & Hamberger (1995) examined the children's experience of numbers when solving elementary addition and subtraction problems, where it was shown that the children experience numbers in three different ways as:

1. *Number words* - numbers are individually numbers or linked to each other in the number sequence and could be used to tell a quantity or amount of set

2. *Countables* - numbers are numbers linked to each other in the number sequence having a specified sequential order and could be used in counting and numerical operations

3. *Structurables* - numbers are connected to each other in a part-whole relation and could be composed and decomposed when performing numerical operations.

Thus, on a group level, connections can be made from the way in which the children verbally express their experience of numbers, to how they experience numbers when solving problems. Children seem to be aware of the function of numbers as number words, countables, and structurables, both in a conversation situation when talking about numbers, and when solving problems. The implications for the teaching process drawn from the study are that children should be given the opportunity to encounter numbers in various contexts and in meaningful problem solving situations and thereby be given the opportunities to gain new experiences and discover different aspects of numbers.

Identical conclusions were drawn from an earlier study of nine year olds conducted by Ahlberg (1992). In this investigation carried out in three different class-rooms the instruction was followed for half a year studying the children's experience of arithmetical problem solving. As mathematical problem solving often is looked upon as using procedures and applying memorised solving methods a variation was introduced in the teaching process, in order to develop the children's flexibility of mind and get them to be inclined to investigate and make hypotheses.

The variation was introduced along three different dimensions - the content of the problems, the means of expressions and in exposing the variation of the children's ideas and thoughts. The children were working with five different types of problems - with a succession from a non mathematical content to problems in which the mathematical content were severely central. In solving the problems the children used different means of expression in that they talked, wrote and performed numerical operations when solving the problems. One main idea for the teaching was that the children should be exposed for one another thoughts and different ideas to be given possibilities to explore new aspects of the problem content. In opening up for this the children interacted in small groups and discussed and argued about the solutions, thereafter bringing some solutions in front of the whole class to give possibilities for all children to relate their own understanding to these ideas. The results of a comparison between an equivalent control group and the participating children unveiled that the instruction had contributed to that the children participating in the study had become better problem solvers.

Regarding the children's experience of the problems it was revealed that some of them have a taken-for-granted orientation while others have an open orientation. Children with a taken-for-granted orientation have a product-intention and are primarily directed towards giving an answer to the problem. They focus on the *thematic content* and the *numbers,* thereby only experiencing two aspects of the problem. They guess or estimate an answer not using any procedure to solve the problems. Other children having an product intention simultaneously experience *thematic content, numbers* and *operations*. Their intention is to use a procedure and carry out calculations to arrive at a solution to the problem without connecting these aspects to the other content of the problem.

On the other hand, children with an open orientation have a process-intention directed towards searching for an answer to a problem. These children are not principally directed towards numbers and operations in the problem but towards all relevant aspects of the problem content in trying to perceive relations and make hypotheses. They simultaneously experience *the thematic content, numbers, operations and different aspects of the thematic content*. In discerning the component parts in the problem and simultaneously experiencing the internal relationship between these different aspects they grasp the whole - the mathematical structure of the problem - and are able to solve the problem (Ahlberg, 1992, p. 245).

Neuman, (1987) has proposed a complex model for the development of arithmetic skills. She interviewed 105 school-starters - children about three months older than these participating in the current study - when solving problems much alike the problems presented in this study. Very briefly, there could be said that in trying to describe children's development of number concepts she outlines three main categories: Prenumerical conceptions, Early numerical conceptions and Numerical conceptions.

In the "Early numerical conceptions" she describes five different conceptions which the children passes on their way to develop conceptions of number; Movement - Fair shares - Names - Estimated - Estimated names.

The "Numerical conceptions" are described in two different main categories: "Counting" and "Structuring". As mentioned earlier, Neuman (1987) stresses that the understanding of the part-whole relation of the numbers is fundamental, if a child is to develop conceptions of numbers and learn arithmetic skills. She makes a distinction between "counting" and "structuring" and signifies that in order to develop cardinality children have to structure the numbers and experience their part-whole relation.

> Besides the "Structuring" conceptions, which have shown themselves to be the main way towards the ten basic concepts, there is also the "Counting" conception which never emerges in these concepts. Since also some of the school starters have occasionally used strategies related to this conception, it is still taken into account in the categories of description, although described as a

"blind alley" which leads to maths difficulties. (Neuman, 1987, p. 209).

The strategies identified within "Counting" are "Renaming" "Counting all/Counting on using two systems" and "Counting back using two systems". When using this strategies the children will end up with difficulties in mathematics according to Neuman (1987, p. 96).

Ekeblad (1996) tentatively agrees that there is a risk that children who adopt an approach to addition and subtraction where problems are solved by unit-by unit counting never come to experience numbers as a whole-to-part relation, which is the fruitful way of approaching basic arithmetic and suggests that the key to the difficulty with number learning may stem from internal discrepancies in the available ways children experience numbers and number learning. She describes three dimensions, each constituted by two contrasting poles, with reference to the children's ways of experiencing numbers in the context of learning. Numbers experienced by children vary in terms of their organisational integration of the poles of structure and sequence, in terms of their presentational nebulousness or definiteness and in terms of whether they were oriented to as phenomenon of authority or as autonomous phenomena.

Taking the point of departure in Neuman´s (1987) research Marton and Booth (1997) describe five qualitatively different ways of experiencing numbers. In "numbers as names" and "numbers as extents" one or the other of the two main aspects of numbers, the ordinal and the cardinal is focused on and is present in awareness, and that does not give sufficient grounds for solving arithmetic problems. In " counted numbers" and "finger numbers" both the ordinal and the cardinal aspects are produced while dealing with problems, in the first separately, in the second case in an integrated manner, and arithmetic problems can be solved. "Number facts" refers to the simultaneous experience of the ordinal and the cardinal aspects as well as the part-whole relation (p. 65).

Exploring Different Aspects of Children and Numbers

In this overview drawn from some of the relevant research on children and numbers, the central aim was not to focus on the different theoretical standpoints taken in the different research perspectives mentioned, but to illuminate what seems to be the most critical and important features within the different views of children's learning of numerosity.

To summarise it could be said that different aspects of how children understand numbers and learn arithmetic are emphasised to varying extent in the models described. This is to be expected in that such complex phenomena as number sense and numerosity developing over a long period of time in the child's interaction with the world probably has not just one origin but several. Consensus must be stated in that none of the views referenced can get in touch with the totality, explaining and describing "the whole" in outlining a general model of how children come to understand numbers. The only possible claims is to describe different themes and aspects of children's learning of numbers. The critical questions then must be what kind of aspects different views can expose and to what extent these clarify the children's learning of numbers. Do the different perspectives in some way complement each other and together clarify more about the children and numbers than just one and does one perspective just eliminate all others? Does the knowledge we get from several approaches contribute to a more complex understanding or are the various views dealing with quite different questions although the issue such as children and learning numbers is the same? The position taken in this study is that these different views must be seen as complementary in that no perspective could be seen as the one and only describing "the whole", but clarifying different inquires of children learning numbers thus - sometimes rather intertwined - in various ways contributing to an understanding of "the whole".

The intention in this report is trying to relate some of the discussed views of children's mathematical understanding and the way in which children understand mathematics in the context of solving different types of elementary arithmetic problems. The aim is not to shed light on children's performance by comparing them to adults or to give a psycho-genetic analysis of spontaneous genesis and the multiple forms it can take on. Instead the aim is to take the perspectives of the children - trying to describe their experience and the richness in their thinking - in the believe that an understanding of children's various ways of experiencing numbers can contribute to the teaching of mathematics to young children.

Chapter 2

AIM AND METHODOLOGY OF THE STUDY

In this investigation an interpretative analysis has formed the foundation for an overall description of how children go about solving simple addition and subtraction problems, and my aim has been to illustrate and describe the different ways in which children experience the meaning of numbers in a problem solving situation. That is, the primary aim of this study is to examine the different ways in which children experience numbers and how these are related to one another.

Underpinning the investigation is the belief that the ways in which children handle numbers simultaneously constitutes their experience of them as well and that by illustrating these differences it is possible to map children's experiences and interpretations of meaning. In trying to capture the complexity of their experience, focus is upon the children's awareness of the situations and the research is directed towards illustrating their experiences through a description of their different strategies of handling numbers. The phrase "ways of handling" is used in connection with numbers or problems to describe how children talk and act when solving problems. It does not, however, denote conscious planning, execution, and control of activity as the term "strategy" most often does (Ahlberg & Csocsán, 1994).

Siegler and Jenkins (1989) define "strategy" as any procedure that is non obligatory and goal directed. They do not stipulate that a "strategy" must be consciously formulated or the product of a conscious or rational choice, and reserve the term "strategies" for procedures that are chosen among. In this study the children do not plan their activity in any observed way nor do they consciously make different strategy choices. The definition of "strategy" used by Siegler and Jenkins is therefore very similar to the way in which the phrase "ways of handling" is used here, but even though "strategy" is the conventional term used in most literature on children's mathematical thinking, the phrase "ways of handling" seems to be more appropriate for this study. The reason for this is that the analyses of children's acting and thinking is not directed towards their actual mental acts, but aims to show how numbers appear to children by providing a description in terms of their own experience.

Method and Theoretical Framework

This study has been conducted within the phenomenographic approach to research and emerges from an experiential perspective. Phenomenography is the empirical study of the differing ways in which people experience, perceive, apprehend, understand, and/or conceptualise various phenomena in, and aspects of, the world around them (Marton, 1981). The aim is to describe experience itself and to look at the world together with other individuals in order to see it as they see it (Marton, 1988). Phenomenography has a non-dualistic ontology in the sense that people's understanding is constituted in the relationship between the individual and the world. The aim is not to try to describe things "as they are" but to characterise how different phenomena appear to people (Marton, 1994).

People interpret and understand the world in different ways according to different experiences, perspectives, and contexts. The phenomenographer seeks to reveal, to analyse, and primarily to describe these in categories which represent relationships between the subjects and the phenomena and encapsulate the distinctly different ways in which the subjects experience this, in the sense of what they see as its meaning, and how they distinguish it from other experiences.

So far, various terms such as "conceptions" or "ways of understanding" have mostly been used to describe the different ways people have of perceiving, understanding, and conceptualising phenomena in the world around them, resulting in categories of description. In this study - like in some earlier phenomenographic studies (Ahlberg, 1992) - this is referred to as "different ways of experiencing".

Marton and Booth (1997) stress that the different terms used in phenomenographic studies should be interpreted in the experiential sense and not in a psychological or physical sense.

> ... descriptions of experience are not psychological and not physical. They are descriptions of the internal relation between persons and phenomena; ways in which persons experience a given phenomenon and ways in which a phenomenon is experienced by persons (p. 122).

The aim of this study is to illuminate and describe *children's various ways of handling numbers and how this is related to their various ways of experiencing the meaning of numbers.* This is an experiential perspective in the sense that the researcher tries to take the point of view of the children and to describe their experienced world. It follows from this that, in the analyses and interpretation of the empirical material, the situation and a child's ways of handling numbers and thinking are looked upon as a unit. This approach was adopted in the afore mentioned study by Ahlberg, (1992) examining nine year olds solving problems in a class-room context. The dynamic nature of the problem solving process was analysed in terms of internal and external frame of reference and the children's awareness was described in three complementary dimensions - direction - reference - movement - revealing the various aspects of the problem that the children experienced simultaneously.

Phenomenographic research studies are interested in how their findings apply to the group or population as a whole rather than the individual. Individuals are seen as the bearers of different ways of experiencing a phenomenon and the description reached is a description of variation on the collective level (Marton & Booth, 1997). In the description of the results of this study the children's experience of numbers on a collective level is outlined, even though the results are also applied and discussed in the light of individual cases.

The subject of the research focused upon are the different ways in which children experience numbers - in the sense of exploring different aspects and possible qualities of the numbers. The findings have been based on two different levels of analyses. The level of how children *handle numbers* is based upon the observations and interpretations of the children's ways of acting and thinking while solving the problems. The level of *experiencing numbers* is made up of these acts and thoughts interpreted and outlined in terms of awareness and experiences - that is, what aspects of numbers children are focally aware of in the problem solving situation.

The Phenomenographic Interview

The main mode of data production in phenomenography is the phenomenographic interview which is a form of interview-as-conversation in the sense Kvale (1996) outlines when defining one form of research interview as:

> ... an interview whose purpose is to obtain descriptions of the life world of the interviewee with respect to interpreting the meaning of the described phenomena (p. 5).

Kvale describes the use of the research interview as a conversation that has a structure and a purpose. It goes beyond the spontaneous exchange of views in everyday conversation and becomes a careful questioning and listening approach with the purpose of obtaining thoroughly tested knowledge. The research interview is not a conversation between equal partners, because the interviewer defines and controls the situation. The topic of the interview is introduced by the researcher, who also critically follows up the answers to the questions.

In the reset reported here all interviews were preceded by a phase where a mutual agreement was reached between the interviewer and the children as to the purpose of the interview, in order to get the children relaxed and enable them to focus on the content rather than on the situation. The interviews were semi-structured in that they had a number of pre-determined problems for exploration, which had been chosen for their ability to penetrate the different ways in which the children experienced numbers, but the interviewer was also sensitive to potentially productive turns in discourse as well as to shifts of focus and moves away from the phenomenon of interest. The interviews were made in a deep and open mode and the interviewer tried hard to understand and respond to the children's acting and thinking by asking accurate "follow up" questions such as: "How do you know that? - How did you figure that out? - Can you tell me how you go about solving the problem? - Can you show me?". She tried to get the children to expound upon and elucidate their answers and explain in more detail how they went about solving the problems.

Questions about Validity and Reliability

The issues of validity and reliability concerning qualitative research such as phenomenography are often discussed and questions as to what extent the results can be relied upon and generalised are regularly posed. To question reliability may be to ask; If another researcher repeated the research would the results arrived at be the same? That is, would the same interpretations and the same descriptions have been made? One way of testing reliability within phenomenographic studies is that a second researcher is asked to study the empirical material in the light of the presented results and thereafter one compares the interpretations (Johansson, Marton, & Svensson, 1985). However, for several reasons this kind of inter-judge reliability is today questioned by many phenomenographers. For example, the interpretations given could be influenced by the researchers' pre-understanding and be unfaithful to the interviewed persons even though they might have a high degree agreement with a co-judge because the experiences might be easily recognised. Further, the researcher's interpretations rest on a much wider experience of the collected data, and even the interview situation, while the co-judges interpretation of an extract from the data can not do it justice.

In the study described in this report another researcher was given access to the interviews, and to the children's various ways of handling and experiencing numbers, and categorised all the problem solving attempts. The purpose of this procedure was to *examine the possibility of multiple interpretations and to explore new conceivable and fruitful interpretations while gaining new perspectives on the empirical material*. The process of interpretation was given depth by undertaking an interactive standpoint, discussing the all the findings, not only from the point of view of the rather few cases - regarding the big amount of trials - in which there was a disagreement on interpretation. Watching selected cases on video-tape several times, the researcher and the co-judge co-operating, discussing, and building up a mutual agreement of interpretation, contributed to the analyses and to the verification and validity of the conclusions.

However, to question validity would be to question the whole study with all its separate stages and what justification there is for presenting the work and making the claims proposed from the results. These questions are intertwined with every aspect of the study and could not be deeply scrutinised in this report and are mentioned only briefly. Kvale, (1996) discusses three different aspects of validity in qualitative interview studies.

Validity as Quality of Craftsmanship - based on the quality of his or her past research in the area, the credibility of the researcher is an important aspect for fellow researchers when ascribing validity to the findings reported.

Communicative Validity - involves testing the validity of knowledge claimed in a dialogue. Valid knowledge is constituted when conflicting knowledge claims are argued in a dialogue: What is a valid observation is decided through the argumentation of the participants in a discourse.

Pragmatic validity - is verification in the literal sense - "to make true". In the pragmatic validation of a knowledge claim, justification is replaced by application, it goes further than communication with a commitment to act on the interpretations.

In a research approach aiming to reveal the meaning in what people say and do, the results can not be looked upon in terms of truth or falsehood. Rather the conclusions should be considered to be a specific way of perceiving the research problem and whether one rejects or approves of this way of interpreting and understanding the problem.

Looking at this study in light of the above validity questions, it can be argued that the researcher in question is familiar with the topics covered in the interviews. The interviews were made in a deep and open mode and the analyses continually questioned the data in the search for convincing interpretations. The results and conclusions of the study should be understood and recognised by those active in the field of study and thereby made communicable to the research community as a whole. The intention has been to follow the guidelines set by Säljö (1988) in validating the conclusions by discussing the internal logic of the children's different ways of experiencing things and to verify the results arrived at by conducting a comparative discussion of the results presented and the results of other research in the field. In this study a great number of extracts and quotes are presented in order to make it possible to judge the validity of the interpretations and to open the findings to scepticism and questioning. The results of the study are discussed in relation to earlier research in the field and presented in such a way that a dialogue can be established. The findings are also discussed with the teaching process in mind and with thought given as to the possible ways they could serve as a guideline for children who are learning numbers.

The Empirical Study

Data for the current study was collected through video-taped interviews with 38 children, 18 girls and 20 boys, in their final year at one of the three different pre-schools participating in the study. At the time of the interviews the children had not received any formal instruction in mathematics. Their average age was 6.9 years. The data production was carried out over a two week period at the end of May and beginning of June 1995.

Interviews

The interviews lasted between 25 to 60 minutes and took place in a separate room at the children's pre-schools. To begin with the interviewer talked with the child about various things in order to create a relaxed and comfortable atmosphere. Thereafter she gave the child some tasks perform, in order to judge the child's mathematical competence. The child was asked to count on the number sequence, telling which number comes before or after another, which number is the biggest, etc., and to show a distinct number of fingers. These tasks are not going to be analysed in this report, but provide a backdrop for the study and sometimes shed light on the various interpretations of the children's way of solving the problems. Then the interviewer asked the child to solve some addition and subtraction problems which were followed by several questions designed to get the child to describe and explain how she/he was thinking. The interviews were carefully transcribed with the intention of providing a rich and distinct description of how each child talked and acted while solving the problems.

Data Analyses

The author initially went through the material collected in order to get an overview of the empirical data. The video-taped interviews provided rich empirical data, including the children's bodily as well as verbal language. The videos and transcripts were analysed consecutively, first by analysing the recordings to get an overall picture of how the children went about solving the given tasks, then the recordings were examined to gain different perspectives on the material and a deeper understanding of each child's thinking.

The different ways in which the children solved the problems revealed both differences and similarities in the children's acting and thinking, and descriptions of how they grasped different aspects of numbers and experienced meaning gradually crystallised. These were explored in more depth by analysing each child and the extracts from every specific task and the different ways in which the children understood the meaning of numbers were related to each other in an integrated and interactive analysis.

The search for the different ways in which children experience numbers - in the sense of being aware of different aspects of numbers - was an exceptionally explorative process and the final findings were successively and gradually defined. The intention was to observe and interpret the way the children thought and acted while solving the problems. These interpretations further served as a basis for trying to reveal the meaning of these acts and thoughts and to expose what aspects of numbers children were focally aware of in the problem solving situation. The aim was also to carefully reveal the way in which a single child goes about solving such problems, taking into account the dynamic aspect of how children's different ways of experiencing numbers evolve.

Tasks

The children were given different kinds of word problems which involved simple addition and subtraction and also included missing addend problems, considered to be rather difficult for this children of this age to solve (Carpenter & Moser, 1982). When solving the problems *the children did not have any manipulative objects to handle.* In one type of problem however, they initially used manipulative objects in solving an introduction question in order to make it easier for them to understand the structure of the problem. These problem solving attempts are not included in the analyses. All the children were given the same tasks to perform, but for some children the tasks within the larger number range were very difficult. When the interviewer realised that these children were not able to give an answer she did not pose any more tasks of this kind within the large number range. Thus the number of problem solving trial varies from child to child, but on an average they tried to solve 14 problems.

The numerical content in the tasks was imbedded in three different contexts in which the semantic content and structure of the problems varied.

Every day Problems

The first kind of task involved situations well known to the children such as getting or losing money, fruit, or crayons (appendix 1). They were asked to solve one addition, one subtraction, and one missing addend problem, one of each kind in the number range 1-5.

> If you have 2 apples and get 3 more, how many do you have then?
>
> If you have 4 pennies[1] and lose 2 of them, how many pennies have you got left?
>
> Imagine that you have 5 pennies in your purse when you go out. When you come home you have only 2 pennies in your purse. How many pennies have you lost?

---

[1] Pennies are used in the text instead of the Swedish word "kronor".

Thereafter the children were presented with problems of the same structure in successively larger number ranges.

Decomposition Problems[2]

These problems were designed to show if the children were able to decompose a specific number into parts and wholes and were presented to the child in the following way. The interviewer initially put 9 buttons and 2 boxes - one white and one black - on the table and asked the child: "Can you count how many buttons there are altogether?" Then she said: "Now I want you to help me to figure out what we can do, if we want to put these 9 buttons into these 2 boxes. How many buttons can we put in each box?"

When the children had given an answer the interviewer asked questions about they had thought and behaved such as "How did you figure that out?", "How did you think now?", "How did you come to think of that?" etc. Even if the child gave a wrong answer these questions were asked. If the child gave a correct answer, the interviewer asked if there was also another way of doing it. The interviewer went on asking for different solutions until the child gave all of the solutions he or she could and was not able to think of another answer. If the child said he/she did not know, the interviewer asked if it would be possible to come up with an answer by counting in another way. The children who were not able to give an answer with nine buttons were presented with the same task, but with only five buttons.

Contextual Problems[3]

The problems were proceeded with a talk about the activities of a baker and a thief who stole buns (appendix 2). To make it easier for the children to understand the content of the problems they were initially shown a plate with a number of paper buns and allowed to solve the first problem of this kind by using these manipulative objects. This way of introducing the problems differs from the other tasks and is the reason why it is called contextual. The problem

---

[2] This is a variant of a task called "The guessing game" used by Neuman (1987).

[3] This type of task is modelled on a computer game used by Ekeblad (1996).

used to introduce the tasks was: "There are 8 buns on the plate. When the baker comes back there are only 2 buns. How many buns has the thief taken?" It was only when solving this very first problem that the children had access to the buns, later they had to arrive at an answer without the help of the objects. After solving the first problem the interviewer gave the children other mathematical problems concerning the story of the baker using increasingly larger number ranges.

Chapter 3

RESULTS

The purpose of the results presented is to describe the relationship between the different ways in which children handle numbers and their experience of meaning, and thereby revealing how children come to understand the cardinality of numbers and learn arithmetic skills. Firstly, the different ways children have of handling numbers are described[1]. Each way of handling numbers is directly followed by an interpretation of what aspects of numbers children experience when using the described way of handling. Thereafter, their main ways of handling the numbers are revisited and further scrutinised in order to show the various ways of handling numbers used by each child. This is followed by a description of the nature of the variation in the ways in which comparison between the different ways in which children handle numbers and their ways of experiencing numbers, summing up the framework by outlining the dynamics of experiencing numbers.

Children's Ways of Handling and Experiencing Numbers

The object of this research is to capture children's experiences and the various ways in which they handle numbers when trying to solve verbal problems. The observations and interpretations of the different ways in which children handle numbers - their actions and thoughts while solving problems - served as a basis for revealing their different ways of experiencing numbers and includes two

1 In order to give a rich and distinct picture of the interpretation of the empirical material a substantial body of quotes are given, thereby elucidating both canonical and non-canonical addition and subtraction problems.

different levels of description. First, a level of *handling the numbers* - usually referred to as strategy use - second, a level of *experiencing the meaning of numbers*. These two levels of description are intertwined and related to each other, but nevertheless, it is quite possible to distinguish between them. There is no single one to one correspondence between children's different ways of handling numbers and their experience of meaning. One specific way of handling numbers does not reflect just one meaning and one way of experiencing the meaning of numbers can be signified by different ways of handling numbers. Five main ways of handling numbers were identified:

WAYS OF HANDLING NUMBERS

- SAYING NUMBERS
- ESTIMATING
- COUNTING
- STRUCTURING
- USING KNOWN FACTS

When using these different ways of handling numbers children's awareness is directed towards various aspects of them. These different aspects of numbers presented in the children's awareness constitute their understanding and consequently they understand the meaning of numbers in qualitatively different ways. The interpretative analysis reveals that when solving verbal problems children experience the meaning of numbers in four different ways:

WAYS OF EXPERIENCING NUMBERS

- NUMBER WORDS
- EXTENTS
- POSITIONS IN SEQUENCE
- COMPOSITE UNITS

Ways of Handling Numbers; Saying Numbers

Sometimes children try to solve the problems by *"saying numbers"*. Of course if children have a solution to a problem they always answer with a number. However, "saying numbers" implies that they give an answer without making any calculation or an estimation of a probable answer, that is, they do not count or make use of procedures for exact quantification, even though they do not know the correct answer. Three different ways of "saying numbers" are identified:

1. Random Number Words
2. Equal Numbers
3. Successive Numbers

Random Number Words

One way of "saying numbers" is to mention a "random number word". This means that children say a number which is not in any way connected to the content of the problem. A child can give these kind of answers while performing *addition* or *subtraction*. Often answers are given very quickly, but sometimes children say a number word after thinking for a rather long time about how to solve the problem. Anna answers rather quickly when she is trying to solve an addition problem in the larger number range.

> I: If you have 13 pennies and get 5 more, how many pennies do you have?
>
> A: 10.
>
> I: Did you use your fingers?
>
> A: No.
>
> I: How did you do it?
>
> A: Thought.
>
> I: Did you think in any special way?
>
> A: Just said it.

"Saying numbers" is found in connection with all the different types of problems presented in this study. The answers can be both inside or outside the number range. For example, Helen is trying to solve a missing addend problem.

> I: The baker has baked buns and they are on the plate. Then he went for a walk and the bun thief came and took 3 buns. When the baker came back there were 2 buns left on the plate. How many had he baked?

K: 9.

I: How did you work it out?

K: Guessed.

The answers are not connected to the content of the problem, instead, the child thinks of a random number while in the problem solving situation. As the children do not use any procedure which they can describe, they then have a difficult time describing their action or thought. Most often they say that they know or guess a number. However, sometimes a child gives an explanation like Emil does, when he is asked to explain his thinking when trying to solve a subtraction problem. He states that the answer is six, because he is six years old.

Equal Numbers

The second way of "saying numbers" is to say "equal numbers" as when the child answers with a number already mentioned in the problem. According to the content and structure of the problems, the "first number" or the "last number" mentioned in the presented problem can be answered, or there can be "equal parts".

First Number

Some children focus on the "first number" mentioned in a problem, when trying to solve all kinds of *addition* and *subtraction* problems. For example, Carina does not use any procedure for solving a missing addend problem, she just focuses on the "first number".

I: If you had 5 pennies in your pocket when you went out and there are 2 left when you get home, how many have you lost?

C: 5.

I: How did you work it out?

C: 'Cos you said 5.

Occasionally, children try to use their fingers, but they do not really know how to go about this and therefore they are not helped by using their fingers to arrive at an answer. Most often they count the first part using their fingers and then give this as an answer. Anna, for example, gives a correct answer to some of the problems by counting on her fingers. However, she had difficulty performing a subtraction problem - despite the help she got by counting on her fingers - and just focused on the "first number".

I: If you have 4 pennies and lose 2, how many do you have left?

A: 4.

I: Can you show me with your fingers? (Counts four fingers on right hand).

I: You had 4 and lose 2? (Still holds up hand with four fingers).

I: How many do you have?

A: 4.

Another child, Petra, never used her fingers when trying to solve the problems. When trying to solve a missing addend problem her awareness was directed towards the first number mentioned in the problem. She thought for quite a while and, like some other children, said she was counting. She probably counted in her head to 8, the first number mentioned in the problem, for she then refers to this as the answer.

I: The baker has baked 8 buns and put them on the plate. Then the bun thief comes and takes some so that when the baker comes back there are only 2 left. How many buns did the bun thief take?

P: (Thinks a while) 8.

I: 8? How did you work it out?

P: I counted cos I can't do take aways yet.

Last Number

Other children focus on the "last number" mentioned in a problem and even this way of giving an answer is used both in trying to solve *addition and subtraction* problems. One of the boys, Martin, knows what to do with his money, but is still not directed towards performing addition in order to arrive at an exact answer.

I: If you have 2 pennies and get 3 more, how many do you have then?

M: I'll buy them all up.

I: Yes, but if you start with 2 pennies and get 3 more how many do you have to buy with?

M: 3.

I: How were you thinking when you worked it out?

M: I don't know.

Another child, Mia, is trying to solve a missing addend problem. She is having trouble understanding the structure of the problem and gives the "last number" as an answer.

I: If you've got 3 pennies but want to buy an ice-cream for 7, how many more pennies do you need?

M: 7.

I: But if you get 3 pennies from me to buy ice-cream and you go to the shop but they say that ice-cream costs 7 pennies, you come home and say, "Ice-cream costs 7 pennies and I don't have enough." How many more pennies do you need?

M: 7.

Equal Parts

When children are solving the decomposition task by trying to divide buttons and place them into two boxes, it is not unusual for them to focus on "equal numbers" and divide the buttons in "equal parts". For instance, David has counted the buttons and knows there are five pieces. Still, when he is trying to solve the problem, he is not directed towards quantification. He mentions "equal numbers" and comes up with an answer far outside the number range.

I: If we have these 5 buttons and are going to put them in these 2 boxes, how should we put them in?

D: You can put them in different ways.

I: Yes, how many are you going to put in there (the black box)?

D: 5.

I: And how many in that one? (points to the white box)

D: ... 5.

Kristin divides a larger amount of buttons. She has counted them and knows there are nine. This knowledge however, does not influence her problem solving. She focuses on the number before the one she has just counted and says "equal numbers", thereby dividing in "equal parts".

I: How many buttons in each box?

K: (She thinks a while) 8.

I: How would you arrange them? How many would you put in the black one?

K: 8.

I: 8. And how many in the white?

K: 8.

I: 8. Is there any other way of doing it?

K: (Shakes her head).

Successive Numbers

The third way of "saying numbers" is to focus on "successive numbers", where children say a number next to one of those mentioned in the given problem by taking one step upwards or downwards on the number sequence. This is what Kristin, (mentioned above) is doing, in addition to focusing on equal parts. This way of handling numbers by going one step up or down on the number sequence is sometimes influenced by the content of the problem. With an addition problem children tend to go upwards, if it is a subtraction problem they either go upwards or downwards. It is not unusual for children to count the number words of one of the parts of the problem aloud or in their head before giving a "successive number" as an answer.

The children may focus on the "first number" or the "last number" mentioned. Eva is trying to solve an *addition* problem by focusing on the last number. She says that she is counting when going one step further on the number sequence.

I: If you have 2 crayons and get 7 from me, how many do you have then?

E: 8

I: How did you work it out?

E: I counted in my head.

I: Did you use your fingers?

E: No, if you say 7, I think its 8 'cos after 7 comes 8.

Some children try to count on the number sequence to arrive at an answer, but they do not know how to go about making an exact quantification. When Hanna is solving a *subtraction* problem she is counting upwards to the number before the last one mentioned in the problem.

> I: If you have 10 candies and you give away 7 of them, how many candies do you have left?
>
> H: (Mumbles) 1, 2, 3, 4, 5, 6 .... 6.
>
> I: How did you do it?
>
> H: Counted to 6.

Another child, Emelie, is also going to perform a *subtraction*. She focuses on the last number mentioned and takes a step upwards on the number sequence.

> I: If you have 10 sweets and eat some so that you only have 3 left, how many did you eat?
>
> E: 4.
>
> I: 4... Now, can you tell me how you worked it out?
>
> E: I just guessed.
>
> I: What made you guess just 4?
>
> E: I don't know.

Petra refers to the first number and then goes backwards on the number sequence when she solves a missing addend problem.

> I: There are 6 buns left on the plate and the baker goes out for a while and the bun thief comes and takes some. When the baker comes back there are only 3 buns left. How many did the bun thief take?
>
> P: (Thinks) 5.
>
> I: 5. How did you get that?
>
> P: I had to think.
>
> I: How did you think?
>
> P: I counted. I counted backwards
>
> I: Backwards from?
>
> P: 6.
>
> I: Yes, but how did you get five?

P: I just did.

Sometimes a child counts one part while simultaneously tapping the table, trying to keep track of the numbers. Dan, for example, while trying to divide nine buttons into the two boxes, touches the table with his index finger while counting one part.

> I: If I put 3 buttons in the black one and the rest in the other, how many will there be in the other one then?
>
> D: (He taps the table with his index finger and counts silently) 8.
>
> I: 8, mm, can you show me how you did it, it all went to quickly for me.
>
> D: Like this: 1, 2, 3, 4, 5, 6, 7, 8 (counts and taps the table at the same time).

Experiencing Numbers as Number Words

When children are "saying numbers" their handling of numbers does not have a specific quantitative significance, and they are not aware of the exact numerosity or the cardinality of the numbers. In the problem solving situation their awareness is directed focally towards the number words in the problem as a referent to any numerosity. When solving a word problem by "saying numbers" they simply answer with a number word unrelated to the precise numerical relations between parts and whole in the problem and they can be said to experience numbers as *Number Words*

When the interviewer reads the problems to the children they listen to the numbers mentioned and simply give one of these as an answer, or they set out from a number in the problem, giving an answer without doing any calculation. This does not mean, however, that the children lack all knowledge of the quantitative aspects of the numbers. In the initial stage of the interviews, when asked to show a defined number of fingers, such as 5, 8 or 3, all the children were able to do this and when asked various questions about numbers, all the children showed an understanding of some aspects of quantification such as "7 is bigger than 2" and "3 is smaller than 5".

From a very early age children come into contact with numbers in various contexts. They learn numbers as names in non quantitative situations, such as knowing they should take bus number five to go to grandmother's. They perhaps know their house number or street number and recognise the numbers of their locker in the swim hall. Numbers are a part of children's daily life and some children - almost as early as they can talk - are able to say a small number sequence, not always starting with one, or including all numbers in the standard sequence, however. By the time children are six years old, most of them are used to counting and some have considerable knowledge about the number sequence (Ahlberg, 1995). Understanding numbers as number words has its roots in children's early experience of numbers. Six year olds have various ideas about the meaning and function of numbers and some ways of understanding counting include saying numbers not linked to one another in the number sequence, such as 3, 10, 6 or 18, or to count on the number sequence (Ahlberg & Hamberger, 1995).

In problem solving situations children recognise the numbers in the given problems and are aware of the fact that they are going to try to solve the problems as well as the fact that the answer should be a number. However, some children do not understand the structure of the problem and do not know any procedure with which to solve it. Therefore, it is quite logical for them to just say numbers such as "random number word", "equal number", or "successive number". In this situation the children's awareness is directed towards one aspect of numbers - "Number Words" which can be used to give an answer to numerical problems.

Experience of Numbers when Handling Numbers "Saying Random Number Words"

The different ways in which a child handles numbers depends upon the content and structure of the problem at hand. It is not unusual for a child to just refer to a *"random number word"* - a number not in any way connected to the content of the problem - in an unfamiliar number range, while in more familiar number ranges making use of a procedure to solve the problem. Ahlberg and Hamberger (1995) found that it is rather common that children who solve problems using a procedure within a smaller, well known number range, do not make use of any procedure but simply give an answer "saying numbers" when trying to solve problems in larger number ranges .

Anna (p 36) does not know how to go about solving addition problems in a larger number range and answers with "random number words". However, when she is performing addition in a smaller - more well known number range - she knows how to arrive at an answer. When the interviewer asks her to solve the problem: "If you have three apples and get two more, how many do you have?," she takes out both of her hands, counts three fingers on one hand, two fingers on the other, and thereafter all five fingers. Thus the content and structure of the problem influence Anna's experience of the numbers. When solving problems in larger number ranges her awareness is directed focally towards numbers as "Number Words", meaning that the numbers do not refer to a precise numerosity.

Experience of Numbers when Handling Numbers "Saying Equal Numbers"

It is rather common that children say *"equal numbers"*. Even older children handle numbers in this way when they do not understand the structure of the problem. In an earlier study conducted by Ahlberg (1992), a nine year old girl, Gunilla, is trying to solve a word problem.

I: In a class with 19 children the teacher gives one math book and 2 writing books to every child. How many books did she give out?

G: How many was it? 19?

I: Mmm. 19.

G: ....

I: Can you work it out?

G: 19.

I: Why?

G: There were 19 children and she had books to all of them so that's 19 books.

Gunilla handles numbers in the same way as some of the six year olds in that she finds the problem difficult and does not use any procedure to solve it. She is primarily directed towards giving an answer to the problem, which Ahlberg (1992) refers to as having a "product intention" and focuses on one of the numbers mentioned in the problem and gives this as an answer. Thus she is only experiencing two aspects of the problem - namely the theme of the problem and the first number mentioned and therefore it is not possible for her to arrive at a correct answer.

The structure of the problem is of great importance to the child's way of handling the numbers. Mia (p 39) is trying to figure out how many more pennies she needs, if she got 3 pennies and want to buy an ice-cream for 7 pennies. She answers with the "last number", however, if the problem had been formulated with 7 as the first number in the problem, perhaps she would have said the "first number". As mentioned earlier, the children "saying numbers" are aware of numbers as numerosity, however, they do not know the exact numerosity. Therefore it is possible that some children focus on the largest number mentioned in the problem, when trying to solve problems dealing with pennies, as they figure that it is worthwhile to have many pennies - because you can then buy a larger ice-cream. It is also plausible that if she had been given the opportunity to use objects - counting with "pennies", for example- she might have been able to handle the numbers in a different way by performing an operation.

Even when children divide buttons and place them into two boxes some of them focus on equal numbers. They are not directed towards the relation of the numbers or exact quantification, but to earlier experiences of sharing in social situations. These experiences may influence children's thinking when trying to solve the decomposition task. When David (p 39) is dividing five buttons and placing them in two boxes his intention is not upon quantification, but initially focuses on equal numbers and suggests five - the number mentioned in the problem - in one of the boxes. He may be aware of the fact that one ought to divide equally when he suggests five in the other box too. However, the main focus of his awareness is to say "equal numbers" and in this situation the numbers do not have any quantitative significance for him and he comes up with an answer outside the number range. Thus, when he tries to solve the presented problem in an interview situation, with no objects to handle, he is experiencing numbers as "Number Words".

Experience of Numbers when Handling Numbers "Saying Successive Numbers"

It is quite common that a child who does not have access to a quantitative procedure for solving a problem will use *"successive numbers"*. Children are familiar with the number sequence, at least in smaller number ranges, from various situations in their daily life and when asked to explain the meaning and function of numbers most six years old say that counting is to count on the number sequence (Ahlberg & Hamberger, 1994). When mentioning successive numbers in the problem solving situation, the children's awareness is directed towards the sequence as words without any specific quantitative meaning. Some children may know that it is possible to solve problems counting on the number sequence; however, they do not know of any procedure to do this. Their earlier experience of numbers - counting on the number sequence - influences the problem solving context and the numbers are experienced as "Number Words" linked to one another in the number sequence without any precise numerosity.

Dan, (p 42) was counting on the number sequence while simultaneously tapping the table, marking a row of "dots". The interpretation of this action is that he counts to eight and is supposed to say the successive number - 9 - and thus to have said all of the numbers. However, there are two boxes and the number 9 is related to the second box, so when saying 8 he has therefore arrived at an answer. *He creates a unit between the fingers, the "dots", and the numbers* in that they are all at once, still he is focally aware of the numbers he says and is therefore experiencing numbers as "Number Words".

As shown, there are several different ways in which children can give an answer to a problem by "saying numbers". The primary characteristic feature of these is that children do not consider the quantitative relations of numbers, rather the function of numbers is simply to be an answer to a numerical problem. This way of approaching numbers has, in the terminology of Ahlberg (1992), been called "product intention" in that children have a "taken-for granted orientation" and simply want to *give* an answer to the problem, in contrast to children approaching numbers with an "open orientation" and adopting "process intention" who want to *search for* an answer.

In this study, however, some of the six years olds can be said to begin searching for an answer, even though they simply answer "saying a number", in the sense that they are trying to count on the number sequence in order to reach an answer. However, they do not know how to go about doing a numerical operation and they end up " saying numbers". Hanna (p 41) counts quietly on the number sequence until she reaches the number next to the second one mentioned in the problem and arrives at an answer by "saying successive number". Numbers do not have any quantitative significance for her and she does not have access to any procedure with which to solve the problem, thus she is experiencing numbers as "Number Words". Dan (mentioned above) tries to solve the problem by counting on the number sequence and by keeping track of the counted numbers by tapping on the table. However, as he focuses on numbers as "Number Words" he does not get any help from these ways of handling numbers and is not able to solve the problem.

Conclusion

When children are handling numbers "saying number words" they know that the number word refers to a numerosity; however, the precise numerosity of this is not clear and they experience numbers as "Number Words". In the problem solving situation the children are aware of numbers as an answer to a numerical problem, however, they do not have any cardinal meaning, neither in the sense that the last number word used can be interpreted as an indicator of the counted collection's numerosity, nor in the sense that cardinal numbers are seriated and included in each other and can be decomposed into unit items.

Ways of Handling Numbers; Estimating

Sometimes children use *"estimating"* in order to give an answer to mathematical problems. In this case they do not count on their fingers or use any other procedure to come to an answer. They refer to a number, but in contrast to children who experience numbers as "Number Words", they refer to a plausible answer close to a correct one. This way of handling numbers is used in *addition* as well as *subtraction*.

It is possible that a child knows of a way to make an exact quantification, but does not make use of this knowledge in the problem solving situation. Instead of using a procedure to solve the problem they just give a probable answer. Mikael, for instance, makes a rather fair estimation when trying to solve an addition problem without counting on his fingers as he sometimes does.

> I: If you have 2 crayons and get 7 more, how many do you have?
>
> M: Mm, 2 first and then seven, mm that's 11.
>
> I: How were you thinking?
>
> M: I'm fast. I don't think, I just say it.

Markus on the other hand, uses different ways to arrive at an answer when solving various problems. Most often he counts or uses known facts and once - in a very large number range - he just said a number word. However, he will sometimes estimate an answer as in this subtraction example where he does not count on his fingers but simply gives a probable answer.

> I: If you had 10 sweets but ate some of them so you only had 3 left, how many have you eaten?
>
> M: 6.
>
> I: How did you do that?
>
> M: Thought.

When the number range is larger, it is not unusual that a child estimates even if she/he uses some procedure for solving the problems in smaller number ranges. One child, Moa, most often uses her fingers to arrive at a correct answer, but when the number range is large she sometimes estimates. This is the way she tries to perform a subtraction by saying the number successive to the last number mentioned in the problem - occasionally giving the correct answer.

> I: If you have 15 apples and give 7 away how many do you have left?
>
> M: 15 apples and give away 7.. 8.
>
> I: Can you explain how you got it?
>
> M: I just guessed.

Yet another child, Erik, uses a lot of different approaches when trying to solve mathematical problems. However, when he solves a missing addend problem within a larger number range he estimates.

> I: If you have 12 pennies and want to buy a paper for 19, how many more do you need?
>
> E: 9
>
> I: How did you get 9?
>
> E: I just said it.
>
> I: You just said it. But why just 9, why not 22?
>
> E: No
>
> I: Why not?
>
> E: No. 9 is more right.

In comparison with other ways of handling numbers, "estimating" is not very common, although children use it in solving all kinds of problems, even the decomposition task. The idea of sharing equal parts can thus influence children's thinking. Emelie counts 9 buttons and suggests that there should be 5 buttons in each box. Christian counts 9 buttons as well and thereafter estimates four buttons in each box. He says equal numbers, however it is not any equal numbers, in that his reasoning shows that have an idea of the numerosity of the numbers.

> I: Why do you want to put 4 there and 4 there?
>
> C: Because that's 9.
>
> I: Is it 9?
>
> C: Yes cos 5 + 5 is 10. and 4+4 is ...
>
> I: But if I put 5 there then?

C: You can't?

I: Why can't I?

C: No you don't have enough buttons.

However, when Christian is questioned about his thinking and given some new tasks to solve, he overcomes the idea that it has to be equal parts and starts to make exact quantifications, arriving at correct answers such as 7 and 2, 8 and 1, 6 and 3, 4 and 5.

Experiencing Numbers as Extents

When children are handling numbers by "estimating" they do not use any procedure to come to an answer, even if they do not know the exact answer. Nevertheless, their close to correct estimations indicate that they have a sense of numerosity and the relationship between numbers even though they do not try to use the numbers for exact quantification. They make a probable estimation and refer to numbers as parts and whole, without focusing on their exact numerosity and seem to experience numbers as *Extents*. However, in contrast to children who experience numbers as "Number Words", they have a more precise understanding of the numerosity of the numbers and do not focus on numbers as single or linked words on the number sequence. The task involves more than simply "saying numbers" and, in realising that there is a relation between the addends and the sum - or in other words a relation between parts and the whole in the problem - they seem to have an understanding of the structure of the problem. They can compare and relate numbers and there is a sense of quantitative relationships in that numbers are constituted of complementary parts. However, they are not directed towards individual number units and it is only by chance that the experience of numbers as "Extents" leads to a correct solution, as shown in the case of Moa above (p 48).

When solving the decomposition task - dividing coins into two groups - some children, like David (p 39) simply focus on equal numbers - saying numbers- without taking into account their quantitative relation. Other children, like Christian (p 49), encounter a dilemma when trying to solve this problem. They have an understanding of the part-whole relation of the numbers in the problem; however, their awareness is mainly directed to the idea of "equal parts". From a very early age children are used to sharing objects- food, sweets, and so on with other people and in these situations it is important to divide equally. Thus the belief in sharing equally is developed from experiences of sharing in social situations and these experiences may influence children's thinking when trying to solve decomposition tasks. However, it is contradictory to an exact quantification of the numerical content in the task. In this situation some children make an estimation and give a probable answer close to the correct one by suggesting two equal numbers. Nevertheless, their awareness can shift and, like Christian, (p 49) they begin to make exact quantifications.

Conclusion

The characteristic feature for experiencing numbers as "Extents" is that numbers have cardinal meaning. In the problem solving situation a child's awareness is not directed towards the numbers in the sense that they are to be decomposed in ideal units, seriated and included in each other, but with the focus on comparing and relating the parts and the whole in the problem without performing an exact quantification. In a problem solving situation the child focuses on the approximate numerosity of the numbers and does not consider the precise numerical relations between parts and whole in the problem.

Ways of Handling Numbers; Counting

Handling numbers by *"counting"* covers the different ways in which children give an answer to a problem by counting on the number sequence to make an exact quantification. Sometimes children count without any keeping track procedure, in other cases they keep track of the numbers in various ways. They count on the number sequence to arrive at an answer in five different ways:

1. **Double Counting**
2. Counting and Tapping
3. Counting and Looking
4. Counting and Listening
5. Finger Counting

Double Counting

When children are *"double counting"* they count on two parallel number sequences without using their fingers, but *use one of the number sequences to keep track of the counted numbers*. "Double counting" was a very uncommon way of handling numbers among the interviewed children. None of the children used "double counting" to perform subtraction. However, two of the children did count upwards on two number sequences when performing addition.

As we see below, Daniel uses "double counting" when he solves four of the presented problems.

> I: Now I have these 9 buttons and I want you to help me. We are going to put them into these two boxes, how many should I have in each?
>
> D: 6
>
> I: You put 6 in one and the rest in the other, how many are there then in the other one?
>
> D: 3
>
> I: Do you remember how you thought about it?
>
> D: Yes, that 7 was 1 and 8 was 2 and 9 was 3.

In larger number ranges, as well as in smaller ones, Daniel counts silently in his head on two number sequences in order to keep track of the numbers and does not use his fingers.

> I: If you have 13 pennies and get 5 more how many do you have then?
>
> D: 18.
>
> I: How were you thinking when you did that?

D: Like 14 as 1.

I: 14 as 1?

D: Yea, and 15 as 2

I: 15 as 2

D: And 16 as 3, 17 as 4 and 18 as 5.

When children count on two parallel number sequences they do not listen to how many numbers they say. They count upwards or downwards on the first sequence and use the number words of the second sequence to keep track of the counted numbers. When they have said all number words one after another in the second part of the problem, they have arrived at an answer, and it is time to stop counting.

Counting and Tapping

It happens - but very rarely - that children use *"counting and tapping"* as a way of handling numbers. They tap a row or a pattern with their index finger on the table while counting - most often they count one by one, marking a row on the table - and in this way *trying to keep track of the counted numbers.*

Mikael is often "saying numbers" while trying to solve the problems. However, once he also uses "counting and tapping" with his index finger on the table.

I: You had 5 pennies in your pocket but when you get home you only have 2 left. How many have you lost?

M: (Counts in a row on the table) 1, 2, 3, 4, 5. I've lost that one, that one, and that one .. 3.

As in "double counting", the child sometimes has a difficult time keeping track of the counted numbers and it is not unusual for them to make mistakes using "counting and tapping". When solving another problem in a larger number range Mikael initially tries to estimate an answer. Thereafter he is "counting and tapping" on the table, but loses track of the numbers and does not arrive at a correct answer.

I: If you have 2 crayons and I give you 7 more crayons, how many crayons will you have altogether?

M: Hmm, 2 first and then 7. Its 11.

I: How did you figure that out?

M: I just said it.

I: But can you figure it out in another way?

M: 10. (Firstly he counts 1, 2 tapping on the table and then he counts on until 10, still tapping).

Counting and Looking

Another way of counting is *"counting and looking"*, where children look at something in their surroundings or visualise objects and use them for counting. This way of handling numbers is a way *to keep track of the counted numbers*. In this study only one boy handled numbers by "counting and looking". When trying to solve an addition problem he looked at a big clock hanging on the wall. The clock had big, black lines instead of numbers and Christian counts on these in order to arrive at an answer.

I: If you have 4 pennies and lose 2, how many have you left?

C: 2.

I: How did you think it out?

C: I checked on the clock by counting the lines.

Counting and Listening

When children handle numbers by *"counting and listening"*, they count *upwards* or *downwards* on the number sequence, listening to how many numbers they say without *any procedure to keep track of the numbers*. Sometimes they make a body movement, such as nodding their head, to follow the counting. Children use "counting and listening" when solving both canonical and non-canonical problems. However, it is mostly used when one of the numbers in the problem is small.

David sometimes gives an answer by "saying numbers", but he also counts to solve the problems, thereby shifting between "finger counting" and "counting and listening". When performing subtraction he counts on his fingers, not always arriving at a correct answer, however. When solving *addition* problems he counts upwards and listens to how many numbers he says. He starts to count the largest part, then he just counts on.

I: If you have 2 crayons and then get 7 more, how many do you have then?

D: 1 2 3 4 5 6 7 .. 8 .. 9.

I: How did you work it out?

D: I counted from 2 to 9.

Daniel solves the same problems by "counting and listening" from the largest one. He starts at 7 and then continues counting.

I: If you have 2 crayons and then get 7 more, how many do you have then?

D. 9.

I: How did you work it out?

D: I turned it round so that I had 7 from the start and then counted 2. 8, 9.

I: Ooh, how did you come across that little trick?

D: At home once.

Dan is "counting and listening" when he solves the decomposition task of trying to divide nine coins and place them into two boxes. He states the first number and continues counting, without keeping track of the numbers, simply listening to how many numbers he has said.

I: If I put 5 in the black box how many do I have left to put in the white one?

D: I think it'll be 6, 7, 8, 9.

I: Yes, how many is that?

D: 4, I think.

Erik handles numbers in the same way. When working on an addition problem he counts from the first number, "counting and listening" to how many numbers there are without keeping track on his fingers.

I: The baker has made some buns and the thief sneaks in and pinches 3 buns, so that when the baker comes back there's only 5 left on the plate. How many had the baker baked?

E: 8.

I: How did you think it out?

E: I counted from 5 to 6, 7, 8.

It is not so very common for children to count *downwards* by listening to the number words when solving *subtraction* problems. However, in a small number range Daniel handles the numbers in this way:

I: If you have 4 pennies and lose 2, how many do you have left?

D. (Counting aloud) 3, 2.

In a larger number range Johan counts *downwards* when solving a missing addend problem. He listens to how many numbers there are without keeping track on his fingers.

I: The baker has baked 14 buns but the thief comes and takes some. When the baker comes back there's 8 buns on the plate, how many did the thief take?

J: 6.

I: How did you know that?

J: I counted backwards from 14.

I: Did you count in your head?

J: Yes.

I: Can you say how?

J: 13, 12, 11, 10, 9, 8.

In another problem Johan is counting *upwards* and listening to a great deal of numbers.

I: So the baker's baked again but when he goes out the thief comes and takes 7 buns and leaves 10 on the plate, how many did the baker make to start with?

J: 17.

I: How did you think this time?

J: I counted forwards. 11, 12, 13, 14, 15, 16, 17.

I: How do you know how many you've counted?

J: I've counted lots of times.

Finger Counting

"Finger counting" covers all the different ways that children use their fingers when trying to make an exact quantification. In all different ways of "finger counting" described in this section *children use their fingers in various ways to keep track of the counted numbers.*

Initially, the children were asked to show different numbers of fingers such as 5, 10, 8, and 3. A few counted the fingers one by one, but most of the children were able to show the number of fingers without counting them. It also turned out that, although some children did not count on their fingers, when solving the presented problems they managed to show the number of fingers they were asked to.

Children use their fingers when solving all sorts of problems and use them more frequently in a bigger number range or when they think the problems are difficult. Sometimes they do not want to reveal that they are counting on their fingers and hide them under the table or in their laps. Occasionally they just make a very small movement or press the fingers against something to keep track of the counted numbers.

The number of varieties of using fingers for counting is immense and a child can go about this in various ways when solving different problems. Children's hands can be turned up or down and they may count on the left, the right, or both hands, starting from the left solving one problem and from the right solving another. The same child may begin to count on the little finger, the thumb, or some other finger when solving different problems. Obviously, it is a complicated thing for some children to learn to count on their fingers. *They have to learn by practice and become familiar in the use of their fingers in arriving at a correct answer to the given problems*. The children used their fingers in three different ways in solving the various problems given in the study, both addition problems and subtraction problems.

1. *Using Fingers; Counting All*
   The child counts the two parts as well as the whole on their fingers, one by one.

2. *Using Fingers; Touching*
   The child touches their fingers or makes a small movement with them. They count the fingers one by one or group them without counting.

3. *Using Fingers; Looking*
   The child counts some part or the whole on their fingers and groups the other part and/or the whole without counting those fingers one by one.

Using Fingers; Counting All

"Counting all" - counting one finger after another - is rather common. Here, one finger represents one number and the child successively connects the fingers to a number. In both addition and subtraction some children count all their fingers one at a time but they tend to solve addition problems by counting on their fingers more successfully than they do subtraction ones.

Cia is performing an *addition* problem, for example. She starts by counting the first part, then she counts the second part, and finally the whole.

> I: If you have 2 apples and get 3 more, how many do you have then?
>
> C: 5
>
> I: 5, how did you work it out?
>
> C: By counting with my fingers.
>
> I: Did you? Can you show me how?
>
> C: I took 2 fingers first. And then I took 3. (She counts 1, 2 on her fingers and then 1, 2, 3 and then all) 1, 2, 3, 4, 5.

When working with *subtraction* problems some children also follow the structure of the problem. They start with the whole and count all the numbers on their fingers and then the two parts, as Niclas does below.

I: If you had 10 sweets but eat some up so you only have 3 left, how many have you eaten?

N: 10 (counting quickly his 10 fingers) 1, 2, 3 so I'd eaten 1, 2, 3, 4, 5, 6, 7 (counts on fingers).

It is not unusual for children to make mistakes when counting on their fingers. Moa is trying very hard below, but does not manage to come up with the correct answer.

I: You have 2 crayons but get 7 more, how many crayons do you have now?

M: 10 (counts quietly on her fingers for quite a long time. She takes up one hand and the other with 2 fingers up and then 2 more)

I: What did you do? Did you count your fingers?

M: (Nods).

I: So you had 7 there and 2 more (pointing at Moa's fingers) and you counted them and got 10.

Quite often, children try to use their fingers, but they do not really know how to go about it and do not get any help from their fingers in finding an answer. Tom is very skilful in counting on the number sequence and counts without any problem beyond 100. Out of 18 problem solving trials, he counts fingers 13 times. However, he only manages to give a correct answer in seven of these trials, because the number range is large or he does not understand the problem. He counts on his fingers in different ways. Sometimes he counts all the numbers, but sometimes he starts from the largest number. Once, when trying to solve a problem in a large number range, he counts all the numbers on his fingers but there are too many for him to keep track of.

I: You have 13 pennies and get 5 more, how many do you now have?

T: Can I use my fingers?

I: You can do it however you like.

T: 22. (Use both hands and counts first to 13 then 5 more, but has a problem keeping track of which fingers he's counted).

Using Fingers; Touching

Children sometimes use their fingers, *"touching"*, simply making a small touch or movement with their fingers to keep track of the counted numbers. They can touch anything, a table, a chair, their knee, whatever, or they may press their fingers together. The children do not look at their fingers and often only make a very small movement. Therefore, this way of handling numbers can be troublesome to identify, because it is not always shown clearly on the video-tapes. Some children expose their fingers very distinctly when solving some of the problems, but solving problems within a small number range, they do not explicitly move or look at their fingers.

Anders, for example, makes only very small movements with his fingers, touching his knee.

> I: The baker has made some buns but the bun thief comes and takes three. When the baker comes back there are 4 left. How many did he bake to start with?
>
> D: 7.
>
> I: How did you think this time?
>
> D: Silently.
>
> I: Yes, you didn't look at your fingers but you moved them a bit?
>
> D: Yes.

Some children do not touch or move their fingers in any observable way. When asked to explain their thinking they say that they were thinking of the fingers or feeling them.

> I: If you have 2 apples and get 3 more how many do you have?
>
> M: 5.
>
> I: How did you work out that it was just 5?
>
> M: I thought.
>
> I: Did you? Did your fingers help you?
>
> M: Mmm, I touched them.

Using Fingers; Looking

Some children count some fingers - one at a time - and *group* some others. They do not count the grouped fingers one by one, but are *looking* at them as a unit referring to a number. When children group some fingers they do not have to use "counting all" - that is, counting all fingers one by one - but just look at the grouped fingers to know how many they are. There are several ways in which children are "using fingers looking". They can look at "one part", "the two parts", "the whole", or "the whole and a part". However, they do not look at the complete part-whole relation of all the quantities in the problem.

Looking at One Part

It is rather common that children count "the whole and one part" but not the other part, as they simply look at their fingers to know how many they are. When performing addition, it is common that children are *counting on by stating the cardinal value for the first part* in the problem and simply count the second part until they reach the whole.

Tobias solves an addition problem in a large number range by stating the first subset. Without counting the second subset, he groups his fingers and looks at the manyness of it. By looking at his fingers he knows that he has been adding the second part without counting it one by one up to 5.

> I: If you have 13 pennies and get 5 more, how many do you have?
>
> T: I had 13 and so you take 5 more fingers. (Starts counting on his thumb) 14 15 16 17 18.

Sofia counts *the largest number first* when solving an *addition* problem. She continues counting from the largest number and by "looking" at the manyness of the small part she knows when she has been adding this without counting it one by one.

> I: If you had 2 crayons but got 7 more how many do you have?
>
> S: 9.
>
> I: How did you think it out?

S: I counted to 7 first. 1, 2, 3, 4, 5, 6, 7. (She counts 7 fingers and then continues) 8, 9 (and sees that that's 2 without counting).

In missing addend problems it is common for children to count from the first number, stating the cardinal value for the first subset, and then counting on until the missing part is exhausted. *In looking at the missing part on their fingers they know how many fingers it is without having to count one by one.*

I: If you have 3 pennies but want to buy an ice-cream for 7 how many more pennies do you need?

N: (Counts 4, 5, 6, 7 on his right hand) 4. (*Sees* that it is 4 fingers without counting them).

When performing *subtraction* some children count the whole and the first part and, by grouping and looking at the second part, they see the manyness. Emil use this way of handling numbers when solving a problem in a smaller number range.

I: If you had 4 pennies but lost 2 how many do you have left?

E: If I had 4 and lost 2 (He counts on his fingers). 1, 2, 3, 4 and lose 1, 2. Just 2. (Sees 2 in the second part without counting).

Even in a larger number range some children look at one part without counting one by one.

I: You have 25 pennies and lose 7, how many do you have left?

A: (Counting downwards on her fingers one at a time) 24, 23, 22, 21, 20, 19, 18 (stops counting after taking away 7).

Looking at Two Parts

Another way of "counting and looking" is "looking at two parts". When handling numbers in this way children count the whole without counting the two parts on their fingers one by one. This way of handling numbers is not very common, Here is an example where Tina is dividing 9 coins among the two boxes.

I: Can you do it any other way?

T: 3 in that and 6 in the other.

I: Can you show me how you worked it out?

T: 3 there (holds up one hand with three fingers without counting them) and then 4, 5, 6, 7, 8, 9. 6 (she sees that it's 6 without counting these fingers again).

Looking at the Whole

In contrast to "looking at the parts" some children count the two parts - one number at a time - and thereafter "look at the whole" without counting. This is not a very common way of handling numbers among the children interviewed, but Ninni uses it.

I: If you had 2 crayons but got 7 more how many would you have?

N: (Counts the fingers one at a time. First 1, 2 followed by 1, 2, 3, 4, 5, 6, 7 and she sees that it is 9) 9.

Looking at One Part and the Whole

Another not so common way of handling numbers is to count one part - one by one - in combination with "looking at one part and the whole". It is not unusual for a child to directly perceive the numbers on one of their hands, knowing there are five without counting.

I: If you have 3 apples and get 2 more, how many do you have then?

M: (Shows first 2 fingers without counting. Counts then 3 on the same hand and says 5 without counting) 5.

Even when solving subtraction problems children may count one part - one by one - and look at the other part and the whole. Eva is counting upwards one by one in the example below, but also looking at one part and the whole on her fingers without counting them.

I: You have 10 sweets but eat some so you only have 3 left. How many have you eaten?

E: 7 (Holds up 3 fingers without counting. Then counts the other 7 and knows that it is 10 altogether without counting).

Experiencing Numbers as Positions in Sequence and as Composite Units

Children are "counting" in a lot of various ways when using the number sequence in order to make an exact quantification. In doing this they experience different aspects of numbers. When counting they are aware of numbers as "Number Words" and as "Extents", however, these aspects are not focused upon in the problem solving situation. Instead, their awareness is directed towards cardinal numbers and two other aspects of numbers in that they experience numbers as *Positions in Sequence* and *Composite Units*.

When experiencing numbers as "Positions in Sequence", the child's awareness is directed towards numbers as individual units in sequences and they seem to experience numerosity as positions on the number sequence. The end points in the counting procedure are pivotal and the child does not focus on the composition of cardinal sets and the whole.

When experiencing numbers as "Composite Units" children focus on cardinal numbers which can be decomposed both in single units and in smaller composite units. Thus, contrary to the experience of numbers as "Positions in Sequence" children focus both on single units and the unification of elements.

Experience of Numbers when Handling Numbers "Double Counting"

In *"double counting"* children count on two number sequences *to keep track of the numbers* and do not use their fingers. They count until they reach a defined number, and then instantly know the answer. The numbers are enumerated counting figures, linked to one another in the number sequence, and have a specified sequential order. When children take one step at a time on two number sequences, they experience numbers as "Positions in Sequence".

This way of handling numbers using "double counting" was very rare among the children interviewed, in contrast to the use of "double counting" among blind children who often use this way of handling numbers when solving word problems. Blind children use a lot of various ways of "double counting", for example, they may count up on one sequence and down on the other when solving missing addend problems. Often they are very skilful at this; however, "double counting" can be very demanding on the memory and if children have to say many pairs of numbers in order to reach an answer, it can prove difficult for them to remember and to keep the order of the counted number words and then they do not arrive at an correct answer (Ahlberg & Csocsán 1994). When children use "double counting" they do not grasp the numerosity of the counted sets, as their awareness is not directed towards the numerosity of the numbers, but towards numbers as "Positions in Sequence."

Experience of Numbers when Handling Numbers "Counting and Tapping"

In *"counting and tapping"* children count on the number sequence and mark a row or a pattern with their index finger on the table while counting. Most often they count one by one, tapping a row or a pattern on the table. In tapping they try to *keep track of the counted numbers by creating "landmarks "* and, in this way, know when to stop counting. Handling numbers by "counting and tapping" gives the child a tactile as well as an hearing and visual perception of the numbers counted. Their awareness is directed towards both the numbers as single units - when counting on the number line - and towards the totality of numbers in the part and whole - when trying to grasp the numerosity of the counted part. In this way they simultaneously experience numbers as "Positions in Sequence" and as "Composite Units".

None of the children in this study "tapped" some part of their face (such as the nose, chin or cheek) in order to keep track of the numbers, which some six year olds are known to do when solving verbal problem (Ahlberg & Hamberger, 1995). Nine year olds have also been known to use their pencils to make dots in a somewhat similar way of "counting and tapping". They draw an appropriate pattern on a paper - or directly on the numerals themselves- where every dot has special spot on the numeral. When adding 4 and 5 a child may set out from 4, counting on until 9, by making five "dots" on the numeral five. By using these "dots" as landmarks the child is able to keep track of the counted numbers and grasp the numerosity of the second set in the problem (Ahlberg, 1992).

However, when children, as in the case of Mikael (p 53), do not have a specific pattern for their tapping which can be recognised and repeated, it is difficult for them to grasp the numerosity of the counted part and quite often they do not manage to make an exact quantification, even though they do experience numbers both as "Positions in Sequence" and as "Composite Units".

Experience of Numbers when Handling Numbers "Counting and Looking"

When *"counting and looking"* children look at objects in their surroundings or visualise them in order to keep track of the counted numbers. They are "counting on", grouping the numbers by "looking", and in this way they experience the numbers' numerosity. When children are "counting and looking" they, as in "counting and tapping", can be directed both towards the numbers as single units - counting one by one- and the totality of numbers - grasping the numerosity of set - looking at different parts or the whole in the problem. In this way they simultaneously experience numbers as "Positions in Sequence" and as "Composite Units".

Although it was very unusual for the children in this study to handle numbers by "counting and looking" without using their fingers, the way of handling numbers by counting various sets in the surroundings or by looking at them or imagining them is described in other literature (Ahlberg, 1992; Neuman, 1987). Even a few nine years old children use "counting and looking" to solve addition and subtraction problems. For example, they might look at some objects not related to the problem but accidentally lying on the table, they might count the flowers on a picture, or they might count some other things in their surroundings they simply happen to look at. However, as in "counting and tapping", it can be difficult for them to keep track of the numbers and grasp the numerosity of the counted set and quite often children do not manage to make an exact quantification (Ahlberg, 1992). Even though children do not grasp the numerosity of the second set, this way of handling numbers indicates that they have an experience of numbers as "Positions in Sequence" and as "Composite Units".

Experience of Numbers when Handling Numbers "Counting and Listening"

When handling numbers by *"counting and listening"* children count upwards or downwards on the number sequence without any obvious way of keeping track of the numbers. *They perceive the numerosity of the numbers through hearing* by counting out loud or listening to their inner speech. By getting a sensuous experience of how many numbers they say, or hear in their head, they grasp the totality of these numbers. This way of experiencing numbers is common when the number range in the problems is small, otherwise it is hard to perceive how many numbers have been said. Johan (p 56), however, counts both upwards and downwards in a larger number range, listening to how many numbers there are without keeping track on his fingers.

"Counting and listening" is a very common way of handling numbers among blind children. As is the case with "double counting", some of the blind children can be very skilful at this and it is not unusual for them to be able to "hear" five or six units (Ahlberg & Csocsán, 1995). Although blind children are generally one or two years behind sighted children in mathematical development, it seems that most blind children overcome these difficulties as they grow older. However, it is still unclear if they come to be aware of numbers as composite units in addition to the very common use of "known facts" in the use of "double counting", or if they begin to "count and listen" and then proceed to group the numbers by "hearing". Most likely the blind children's use of language is intertwined with their development of number concepts and as Ahlberg and Csocsán (1995) have shown *touching* is also of great importance for the growth of their understanding of numbers.

When solving problems by "counting and listening", the child's awareness is directed both towards the numbers on the number sequence - in that they count one by one - and the totality of numbers in the sets they are saying - in that they grasp the numerosity of these numbers - and they simultaneously experience numbers as "Positions in Sequence" and as "Composite Units".

Experience of Numbers when Handling Numbers "Finger Counting "

"Finger counting" is the most common way of handling numbers in arithmetic among the six year old children, as shown in earlier studies (Ahlberg & Hamberger, 1995). In solving all of the various problems given in the study, children are *creating a unit between the number words and the fingers* by using their fingers in three different ways for performing both addition and subtraction problems.

One way of "finger counting" is *"using fingers; counting all"*- where children count one finger at a time and keep track of *all* the numbers in a sequentially order. This is similar to the "direct modelling strategy" - counting all objects - outlined by Carpenter and Moser (1984). The child starts counting the first set, followed by the second set, and thereafter counts all. In this study the children did not have any objects to count besides their fingers and - in the interpretation of this study - they are counting their fingers as objects. In doing this they use their fingers as references without distinguishing between fingers and numbers but creating a unit "a oneness" between the number word and the fingers.

A second way of "finger counting" is *"using fingers; touching"* where the children do not look at their fingers, and sometimes do not even make a movement or touch them, but still have a sense of them. The children create a fusion - a unity - between the numbers and the fingers. They do not distinguish between numbers and fingers and the numbers become bodily anchored. When trying to explain how they go about solving the problems, the children sometimes refer to their fingers by saying that they think about them or feel them. Whether a child is counting all numbers or grouping when "using fingers touching" is not obvious. Most likely they handle the numbers in the same way as they do when "using fingers looking", which means that they are occasionally counting all numbers, but also that they sometimes count some numbers one by one and group some others by feeling the numerosity of this set of fingers.

A third way of "finger counting" is *"using fingers; looking"* where children count some numbers one at a time and group some others, seeing the numerosity without counting one by one. This way of handling numbers differs from the "counting on strategies" - "counting on from first" and "counting on from biggest" - described by Carpenter and Moser (1984), in that the children's sensuous experience of the numbers is elucidated. When using fingers in "counting and looking" children are able to perceive the numerosity of the first part - or the second part or the whole - in the problem without counting unit by unit. They have a sensuous experience of the numbers by *"seeing"* the numerosity of the grouped numbers and therefore know when to stop counting. When they count "the first" or "the largest" or start from the first or the largest - holding this in their mind - the interpretation is that they are *subitizing* and *immediately grasp the numerosity* of the part grouped on the fingers. In cases where the number range is large children might group the fingers and get a sensuous experience of numerosity by seeing the parts.

Neuman (1987) suggests that children give each finger a specific "name" and that they are "renaming" their fingers, in the sense that they count the same fingers twice, but with other "names". In this study the term "name" is not used, in coherence with Wynn (1992) who outlines that children at the age of two and three know that the counting words refer to a numerosity, though they do not yet know to which numerosity each word refers. The findings from this interpretation of the empirical material is that children do not give their fingers specific "names", but *create an experienced unity* between the number word and one finger or a specific group of fingers. Neither is the use of the terms "renaming" and "counting on/ back using two systems" used in the analyses of this study. The concepts signified by "renaming" and "counting on/all" in the terminology of Neuman (1987, p 213-218) are referred to in different ways in this study namely, *saying successive numbers, counting all* or *using fingers looking.*

When interpreting the findings of this study, we need to be aware that children using the three different ways of "finger counting" described above - *counting all, using fingers touching, using fingers looking,* - are aware of both numbers as individual units and composite units, and that when using fingers counting they simultaneously experience two aspects of numbers - "Positions in Sequence" and "Composite Units".

Conclusion

Six to seven year olds "count" in many various ways and do have an understanding of the cardinality of numbers even though they have not made a fusion of cardinality and ordinality in the Piagetian sense. They experience that numbers can be quantified and that the last number word mentioned signifies the whole set. However, all children do not experience the meaning of numbers in quite the same way when using different ways of counting. When they use *double counting* they experience only one aspect of numbers, that is numbers as "Positions in Sequence". On the other hand while using *counting and tapping, counting and looking, counting and listening,* and *finger counting* they experience two aspects, namely numbers as "Positions in Sequence" and "Composite Units" simultaneously.

Ways of Handling Numbers; Structuring

When children handle numbers by *structuring* they do not count on the number sequence in order to keep track of the numbers, but rather structure the numbers in the problem in parts and the whole in order to arrive at an answer. There are two qualitatively different ways in which the children use structuring:

1. Structuring Numbers; Seeing
2. Structuring Numbers; Using Derived Facts

Structuring Numbers; Seeing

The child does not use their fingers to count on. Instead they structure their fingers in a part-whole relationship, thus grasping the numerosity of the numbers by seeing. This way of handling numbers is used when children perform addition as well as subtraction; however, it is rather uncommon among six year olds.

Arne structures his fingers into parts and the whole when solving an *addition* problem below.

I: If you have 2 apples and get 3 more how many do you have then?

A: 5. I looked at my fingers. Took 2 and 3. That's 5. (without counting his fingers).

Moa just looks at her fingers in order to arrive at an answer when performing *subtraction*.

I: Moa, if you have 4 pennies but lose 2 how many do you have left?

M: 2. (shows 4 fingers and bends down 2 of them).

Tom usually counts one at a time on his fingers. However, when solving one of the problems, he does not have to count his fingers in order to grasp the numerosity.

I: The baker has made 9 buns but the thief has come and taken some so that when the baker comes back there's only 5 left on the plate. How many did the thief take?

T: He must have taken 4 buns.

I: Can you tell me how you thought this through?

T: I looked at 5 fingers first. (Shows five fingers on each hand and bends down one thumb).

Structuring Numbers; Using Derived Facts

When using derived facts children group numbers without counting their fingers. They do this in two different ways:

1. Deducing an Answer from a Known Fact
2. Grouping Numbers as a basis for further addition/subtraction combinations.

Deducing an Answer from a Known Fact

When deducing the answer children set out from a known number fact and use this as a starting point for performing numerical operations.

Daniel knows a "double" which helps him to solve an addition problem.

> I: If you have 2 apples and get 3 more, how many do you have then?
>
> D: 5
>
> I: Can you remember how you thought about it?
>
> D: Yes
>
> I: How?
>
> D: I know that 2+2 is 4 and that then 2+3 is 5.

Grouping Numbers

Sometimes children group the numbers in parts and use these as a basis for further counting when making further addition/ subtraction combinations.

Johan is performing addition in a larger number range and groups the numbers in subsets in this way when arriving at an answer.

> I: You have 16 books and get 8. How many do you then have?
>
> J: 24
>
> I: Can you explain?
>
> J: I thought from 16 to 20 to 24.

Johan groups the numbers even when performing subtraction. He knows how to decompose the number seven and makes several subtraction operations.

> I If you have 15 pennies and lose 7, how many do you have then?
>
> J 8.
>
> I: How can you be sure?
>
> J: I thought 3 and 3 and 1.
>
> I: What is 3, 3, and 1?
>
> J: 15, 12, 9, 8.

Experiencing Numbers as Composite Units

When children are "structuring" they perceive the part-whole relationship of the numbers in the problem and seem to experience numbers as *"Composite Units"*. The numbers have a cardinal meaning, they are related to one another, and can be composed and decomposed. Also, when "structuring", children are aware of numbers as "Number Words" and "Extents" and "Positions in Sequence"; however, these aspects are not focused upon in the problem solving situation. When solving a problem children using "structuring" are aware of numbers in the parts as well as the whole as composite units, in contrast to children using "counting" who do not grasp the complete part-whole relation of the numbers as composite units.

One way to handle numbers by "structuring" is by "seeing", as when using one's fingers simultaneously to grasp the parts and the whole-ness of the numbers in the problem. In this case children do not count their fingers, but immediately experience the numerosity of the numbers. The visual sense is also used when children are handling numbers by "counting and looking". The difference between the two ways of using the visual sense - the one used in counting and the one used in structuring - is that when children are "counting and looking" they include and combine their immediate grasp of a group of numbers with different ways of counting on the number sequence. They do not grasp the complete part-whole relationship of the numbers in the problem through "looking" as do the children using "seeing". When using "seeing" children do not count in any way, but rather grasp the manyness of numbers entirely through looking at their fingers, and in this way gaining a sensuous experience of the numbers and perceiving their part-whole relationship.

Handling numbers by "counting and looking" and "seeing" is also used by hearing impaired children when solving word problems. Perception of numerosity seems to play an important role in their genuine understanding of cardinal numbers, even though numbers were not perceived through "seeing" to a great extent by the hearing impaired (Frostad & Ahlberg, 1996).

Another way to structure is to "use derived facts" - deducing or grouping the numbers by experiencing their part-whole relation. When children group numbers without counting one by one, they experience the numerosity of the numbers. Johan (p 72) is grouping the number 7 in parts of 3 - 3 - 1, and in this way experiencing the seven-ness of the number in the sense that the number can be composed and decomposed.

According to Fuson (1992), a genuine understanding of cardinality means that children can decompose the cardinal number into ideal unit items, can compare and relate cardinal numbers, and have cardinal numbers embedded within the number word sequence, so that they are seriated and included within each other. When children are structuring they have this understanding of cardinality. In problem solving situations they primarily focus upon numbers as a part-whole relation and their awareness is directed towards the numbers to be composed and decomposed.

Conclusion

When handling numbers by "structuring", children grasp the numerosity of the numbers in the problem in parts and whole-ness. They are aware of numbers as "Number Words", as "Extents", and as "Positions in Sequence", although in the problem solving situation these aspects of numbers are not focused upon. The primary focus of their awareness is upon numbers as "Composite Units", meaning that children have a genuine understanding of cardinal numbers in the sense that the last number word used can be interpreted as an indicator of the counted collection's numerosity, and that cardinal numbers are seriated and included in each other and can be decomposed into unit items.

Ways of Handling Numbers; Using Known Facts

Children *"using known facts"* do not count or perform any operation in order to arrive at an answer, but simply give an answer directly. Children primarily know the answers to the problems in the number range up to five. It is rather common that they know that two and two equals four and also that two and three equals five.

We see that Petra is "saying numbers" when trying to solve most of the problems and she never uses her fingers. When she tries to solve five minus two, she suggests the answer four - a "successive number". But when solving an addition problem, she knows that two and three equals five.

> I: If you have 2 apples and get 3 more how many do you have?
>
> P: 5.
>
> I: How did you get it?
>
> P: I did a plus.

Johan knows more number facts - both addition and subtraction ones- than the other children. When solving the problems he uses a lot of different ways of handling numbers in a larger number range, including both "structuring" and "counting". When using "counting" he does not count on his fingers, but handles numbers by "counting and listening". Still, in the number range up to ten, he knows the number facts.

> I: If you have 2 crayons and get 7 more, how many do you have then?
>
> J: 9
>
> I: Can you explain.
>
> J: 7+2 is 9.

Children mostly "use known facts" when performing addition in small number ranges, but some children, like Tina, know some subtraction facts too.

> I: If you have 4 pennies and lose 2, how many do you have left?
>
> T: 2
>
> I: How did you do it?
>
> T: 4 minus 2 is 2.

Experiencing Numbers as Number Words and as Composite Units

Children "using known facts" experience numbers as "Number Words" and as "Composite Units", but are still aware of numbers as "Extents" and "Positions in Sequence", even though these are not focused upon in the problem solving situation.

There seem to be two different ways of "using known facts". Firstly there are isolated, rote learned number facts - number words - for example, when a very small child know a "double" by heart, such as two and two equals four.

Secondly, there are "experienced known facts" - in that sense that children know how to compose and decompose the known fact. These experienced number facts are learned from handling numbers in different ways in every day life. In using their senses to gain a simultaneous experience of different aspects of numbers through "counting and touching", "counting and looking", and "counting and listening", children come to understand that numbers can be composed and decomposed, consequently experiencing them as "Composite Units". These different ways of using "known facts" are also identified among blind children (Ahlberg & Csocsán, 1995).

Petra (p 75), mentioned above, "says numbers " when trying to solve most of the problems as when she suggests that the answer to a problem concerning five minus two is four - a "successive number". But she knows one number fact by heart, that two and three equals five. However, she is not aware of the relationship between the whole and the parts of the numbers, as they are just a rote learned number fact. If a child's knowledge does not include an awareness of the part-whole relation of the numbers in the problem, then they experience numbers as "Number Words" used to solve an numerical problem and not as "Composite Units". However, it is very rare in this study that a child uses a "known fact" and is not aware of the part - whole relation of the numbers.

Most often children "use known facts" within the number range of five, as with Tina (p 75), who uses number facts within the number range of five, but when she is solving problems with larger numbers she uses "counting". Some children know number facts in larger number ranges, such as Johan (p 75), who knows all the number facts within the range of ten. Solving problems within larger number ranges he handles numbers using "structuring" or "counting". He does not use finger counting when solving problems in wider number ranges as some of the other children knowing number facts within smaller number ranges do.

Conclusion

When children are using "known facts" as "Number Words" the numbers do not have a quantitative significance and children are not aware of the manyness or the part-whole relationship of the numbers. On the other hand, using "known facts" as "experienced number facts" the children do experience numerosity. Through the simultaneous experience of different aspects of numbers, sensuously perceived through "counting" or "structuring", children have come to understand that numbers can be composed and decomposed, consequently experiencing numbers as "Composite Units".

Children's Ways of Handling Numbers Revisited

There is a wide variety of different ways in which children handle numbers and when solving problems a child might shift between many of these various ways. However, interpretation of the child's way of handling numbers is related to how a child goes about solving all of the problems, and the answers are seen in the wider perspective of the whole interview. Most of the ways of handling numbers are illustrated very clearly and there is no doubt about how children go about solving the tasks. Still, it is difficult to count frequencies.

One reason for this is that only 38 children were interviewed so that even though the number of problem solving trials was immense, it could be that the children interviewed preferred one way of handling numbers over another. In comparison to other empirical material of the same kind analysed by the author-(Ahlberg, 1994; 1995; Ahlberg & Hamberger, 1995) - it seems that on a group level the children were not as experienced with numbers as the children in the earlier studies. However, an interpretation and categorisation of the different ways of handling numbers shows a tendency which can be of interest. This is illustrated in Table 1 below. In despite of the exemplary character of the interviews, 12 out of the 520 problem solving attempts could not be categorised, because there were no observable signs of how the children went about solving the problems, and they did not explain their thinking behind the problem solving. These are not included in the overview.

Table 1. Frequencies of the identified ways of handling numbers in the different tasks.

	TASKS			
Ways of Handling Numbers	Every day[2]	Decom-position[3]	Con-textual[4]	Frequen-cies n= 520
SAYING NUMBERS				Σ 168
Random Number Words	5	17	16	38
Equal Numbers	21	36	8	65
Successive Numbers	22	30	13	65
ESTIMATING	16	2	3	Σ21
COUNTING				Σ274
Double Counting	4	1	1	6
Counting and Tapping	4	1	3	8
Counting and Looking	1	-	-	1
Counting and Listening	23	14	25	62
Finger Counting				Σ 197
Using fingers; Counting all	43	10	22	75
Using fingers; Looking	64	4	28	96
Using fingers; Touching	11	2	13	26
STRUCTURING				Σ14
Seeing	3	-	1	4
Using derived facts	3	5	2	10
USING KNOWN FACTS	19	5	19	Σ 43

2 The tasks dealing with problems involving everyday situations as getting or losing money, fruit etc.

3 The tasks involving the decomposition of nine or five buttons

4 The tasks concerning a thief steeling some of the buns a baker had baked.

As shown in Table 1 "saying numbers" is a very common way of handling numbers used in all kind of problems by children and by far the most frequent way of trying to solve the decomposition tasks. Among the different ways of "saying numbers", the most frequent is to say "successive numbers", even though "equal numbers" is very common in the decomposition tasks. Probably, the belief that children have that one ought to divide equally influences their thinking and directs them to give an answer with equal or successive numbers. "Saying numbers" is not as frequent in the contextual problems as in the other problem types. One reason for this might be that the children were given the opportunity to handle some manipulative items when solving an introductory problem in order to easier understand the structure and content of the problem. (It should be pointed out that this task is not included in the analyses of the children's ways of handling and experiencing numbers). Through handling the objects they might have been directed towards the precise magnitudes of the numbers involved in tasks.

In comparison with "saying numbers", the way of handling numbers by "estimating" is not very frequent and is mostly used when solving every day problems. One interpretation of the more frequent use of estimating when solving every day problems might be that numbers appear in situations such as this in some children's every day lives and therefore they are more experienced with the content of these problems and are more likely to make a probable estimation.

The most common way of handling numbers is by "counting" which is used for all kind of problems. However, it is not as frequent in the decomposition problems as it is in the other tasks. There is a wide variety of different ways of "counting". "Double counting", "counting and looking", and "counting and tapping" are very rare. "Counting and listening" is rather common, but definitely the most frequent way to solve the problems is by "finger counting".

It is sometimes hard to interpret and distinguish children's different ways of "counting" when they use their fingers. Sometimes children are handling numbers by "counting and looking", but when asked to explain how they go about their task, they may use "counting all", because they want to explain very explicitly.

Therefore, it is possible that "counting and looking" is an even more frequent way of handling numbers than shown in Table 3.

Structuring numbers by "seeing" or by using "derived facts" is not a very common way of handling numbers. "Using known facts" is somewhat more common. "Structuring" and "using known facts" are almost exceptionally used in the smaller number ranges.

Children's Main Ways of Handling Numbers

There is much variation as to how many children use the different main ways of handling numbers. As shown in Table 2, the frequencies of the problem solving trials correspond to the number of children using the different main ways of handling numbers at least once during the interview.

Table 2. Number of children using the different ways of handling numbers at least once.

Saying Numbers	Estimating	Counting	Structuring	Using Known Facts
32	14	33	6	14

Only five children of the 38 have not been categorised as using the "counting" way of handling numbers at least once in the interviews. Also "saying numbers" is used by most of the children, particularly in the absence of a known procedure for making an exact quantification. Some of the children who use other ways of handling numbers in smaller number ranges, such as "counting", "structuring" or "known facts", are found to be "saying numbers" in the larger number ranges.

More than one third of the children are shown to be "estimating" at least once. As when "saying numbers", however, there are some children who have access to procedures for quantification in small number ranges, but who in larger number ranges (or when they feel tired), do not make use of this knowledge.

"Structuring" is a rather rare way of handling numbers, and is only used by six children. However, in spite of the fact that the children have not had any formal training in mathematics, many of them are aware of numbers as composite units and more than one third of them are able to give one or more known number facts when solving the problems, most frequently within the number range of five.

Handling Numbers in More Than One Way

Most children use more than one way of handling numbers, being influenced by the structure and content of the problems. Table 3 shows the frequency of each of the many different ways of handling numbers a child uses.

Table 3. Frequencies of the children's use of different ways of handling numbers.

Number of different ways of handling	1	2	3	4	5
Number of children	2	17	11	8	-

Very few children use only one way of handling numbers when trying to solve the presented problems. Of these, one child uses "saying numbers" to try to solve all of the problems posed to her, whilst another child only use counting. Most of the children use two main ways of handling numbers, but it is also common to use three ways and even four different ways of handling numbers have been found . However, none of the children use all of the five different ways. The children shift between the different ways of handling numbers when solving various problems. The most common shift is between "counting" and "saying numbers" (11 children), but it is also common to vary between "saying numbers", "estimating", and "counting" (6 children); or "saying numbers", "counting", and "using known facts" (6 children).

"Saying numbers" is combined with all of the other ways in different combinations. However, there is one child, Andreas, who does not solve the problems in any other way. The number of correct answers to the tasks varies from 0 to 25. Andreas does not have any correct answers. Johan is the child who manages to solve the most problems. He does not use "saying numbers", but uses all of the other four ways of solving different problems. When he is "estimating", he does so in a very large number range and comes very close to a correct answer.

There are only five children out of the 38 who do not handle numbers by "counting". Among the children who do not use their fingers are the ones who use "saying numbers" or "estimating", but there are also some children who make exact quantifications, never once using their fingers.

Ninni tries to solve all the problems presented to her by using "finger counting". When the number range is too large for her she does not "estimate", as some other children do, but states that she does not know how to arrive at an answer. She manages to solve 18 of the problems by using "finger counting". When solving some of the problems she is "counting all" in the sense that she is counting all numbers on the fingers one at a time and with other problems she is "looking" in the sense that she counts some numbers and groups some others.

14 children gave a "known fact" as an answer one or several times. The "known facts" they use are mostly in the number range from 1 to 5, such as 2+3, 4-2, and 5-2. Fairly often the children also know some "doubles" such as 2+2, 3+3, and 4+4.

To summarise the findings, most of the children prove to be flexible in their thinking and are able to make use of several different ways of solving the problems. For example, Daniel handles numbers in several various ways. He uses "estimating" on one occasion, but mostly he uses "counting" in different ways, alternating between "counting and listening", "double counting", and "finger counting". In addition he also uses other ways to handle the numbers such as "derived facts" and "known facts".

There are only two children who do not shift between different ways of handling numbers when solving problems. To be able to use different strategies or ways of solving the problem and shift between them, according to the content and structure of the problem and the premises of the tasks, is seen as a very important component in the learning of arithmetic skills (Carpenter & Moser,

1984; Siegler & Jenkins, 1991; Siegler & Crowley, 1991). If children get into difficulties when trying to solve a problem, because they do not remember the "Known facts", the simultaneous experience of numbers as "Number Words" and "Extents" and "Positions in Sequence" and "Composite Units" give them opportunity to use different ways of handling numbers, and some children may even be aware of a repertoire of ways to handle the numbers.

The findings show that most of the six year olds are flexible and prior to receiving any formal instruction, use many different ways of handling numbers, even though they do not always try the most efficient way with which to solve the problem.

Children's Ways of Experiencing Numbers Revisited

This project has illustrated some of the aspects of how children grasp numbers and the way in which their different ways of experiencing numbers are related to one another. The main result shows that there are distinctly different qualities in children's experiences and that children are simultaneously aware of various aspects of numbers. In the following description we will deal with children's experience of numbers on a collective level.

Describing the experience of numbers on a collective level is to describe the nature of the variation in the children's different ways of perceiving, understanding, and experiencing numbers. This description is not restricted to a child as an individual, instead a single child is looked upon as the bearer of an experience or understanding (Marton, 1981). The descriptions therefore, are phrased in terms of ways of experiencing and show that children experience numbers in four different ways: as Number Words, as Extents, as Positions in Sequence, and as Composite Units.

The Nature of the Variation

There is not a one-to-one relationship between ways of handling numbers and how numbers are experienced by children. Even though a certain way of handling numbers on a general level may guide the children's awareness of numbers. When using the same way of handling numbers, children can grasp different aspects of the numbers and experience the meaning of numbers in different ways. The relationship between the handling of numbers and the meaning of numbers is shown in Table 4. In problem solving situations some meanings are *focally presented in the children's awareness* (showed in white) while others are not (showed in grey).

Table 4. *The relationship between ways of handling and ways of experiencing*

	WAYS OF EXPERIENCING NUMBERS			
Ways of Handling Numbers	**Number Words**	**Extents**	**Positions in Sequence**	**Composite Units**
SAYING NUMBERS				
Random Number Words	●			
Equal Numbers	●			
Successive Numbers	●			
ESTIMATING	●	●		
COUNTING				
Double Counting	●		●	
Counting and Tapping	●	●	●	●
Counting and Looking	●	●	●	●
Counting and Listening	●	●	●	●
Finger Counting				
Using Fingers; Counting All	●	●	●	●
Using Fingers; Touching	●	●	●	●
Using Fingers; Looking	●	●	●	●
STRUCTURING				
Seeing	●	●	●	●
Using Derived Facts	●	●	●	●
USING KNOWN FACTS	●	●	●	●

On their way to learning to grasp numerosity children handle and experience numbers in various ways. When they handle numbers by "saying numbers" and "estimating", they give an answer to a problem without the use of procedures, even though they do not "know" the exact answer. By handling numbers in these ways children experience numbers as "Number Words" or "Extents" and do not perceive the numbers through a sensuous experience of the numerosity.

When children are "counting" they can handle the numbers by using a mental procedure through which to arrive at an answer, but they may also use their senses when solving the problems. In "double counting" they count on two number lines until they arrive at an answer by focusing on the end points in the procedure. The numbers are individual units in sequences and children experience numbers as "Positions in Sequence".

In "counting and tapping", "counting and looking", "counting and listening", and "finger counting", children have an experience of numbers as "Positions in Sequence" and "Composite Units". They count on the number sequence and simultaneously have a sensuous experience of the numbers. When children are "counting and listening", they only grasp the numbers through their sense of hearing, but most often sensuous experiences of numbers are intertwined, as in handling numbers by "counting and tapping" where children are listening to the number words and touching the table and getting a tactile sense and perhaps even a visual experience of the numbers at the same time. In "finger counting" in particular, children have access to several senses; however, one can be focused upon, such as the visual sense in "seeing" or the tactile sense in "counting and tapping".

When using "structuring" children can also have a sensuous experience in "seeing" parts and wholes of the numbers. However, structuring can also be purely mental, with the focus on relationships between sets - as when using derived facts. When children say "known facts" they know the answer and do not have a sensuous experience or make use of any procedure with which to solve the problem.

The Sensuous and Simultaneous Experience of Numbers

The findings have been based on two different levels of analyses. What then is the difference between ways of handling and ways of experiencing, and is it really possible to distinguish between the two levels when, as declared, the two levels of description are intertwined and related to one another? The result of the interpretation of the level of *handling numbers* - based on the observations and interpretations of the children's types of action and thought while solving the problems - forms a description of the different ways in which children handle numbers, using their senses in trying to grasp and solve the problems. The result of the level of *experiencing numbers* are those acts and thoughts interpreted and outlined in terms of awareness and experience. That is to say, what aspects of numbers children are focally aware of in problem solving situations.

Analysis shows that children handle and experience numbers in various ways and that different aspects of numbers are focally presented in children's awareness at these times. Also shown is that on the way to understanding numbers sensuous experience is of great importance as in touching, seeing and hearing when trying to solve the problems. "Counting" and, above all, "finger counting", are the most frequent ways of handling numbers among six to seven year old children and may generate sequential thinking rather than forming an understanding of numbers as composite units. Still, even when children are "counting" they go about this in different ways - from "double counting" to using their senses in different ways by "tapping", "looking", "listening", and "finger counting". *When children count and group numbers their sensuous experience enables them to grasp the numerosity of numbers as composite units.*

Experiencing Numbers by Keeping Track and Marking

In order to arrive at an correct answer when solving arithmetic problems children have to capture the numerosity of the numbers through "marking" the numbers either through using a procedure or perceiving them through their senses. When children are "counting" they use several different ways with which to "mark" the numbers. Sometimes they use a procedure, as in "double counting", experiencing numbers as "Positions in Sequences". The first and last counted number on the number sequence then serve as "landmarks" to keep track of the numerosity. When children are "using fingers; counting all" the addends and the sum (or parts and the whole) in the problem serve as "landmarks" in the creation of a unit between one number word and one finger.

It is also a characteristic feature of children's different ways of handling numbers when experiencing numbers as "Positions in Sequences" and as "Composite Units" that sometimes they use their senses to try and keep track of the numbers. When children are "counting and tapping" and "counting and looking" they "mark" the numbers by creating "landmarks" with which to capture the numerosity of the counted set.

When children are "using fingers; looking" they *group* the fingers and get a sensuous experience of the numbers by *seeing* the numerosity. The fingers and numbers form a unity and the child is able to grasp the numerosity through subitizing by grasping the numerosity of the grouped fingers without counting one by one, but through a sensuous experience.

However, when handling numbers by "counting and listening" children do not have any observed way of "keeping track" or "marking" the numbers. They simply *hear* how many numbers they have said to arrive at an answer. Also, when handling numbers using "structuring; using seeing" the children gain a sensuous experience of numbers by grouping their fingers and simultaneously grasping the parts and the whole of the numbers. This is not the case, however, when they are "structuring; using derived facts". They do not then have this sensuous experience, but "mark" the numbers through grouping them by doing several addition or subtraction operations.

The conclusion drawn from these findings is that many children use their sensuous experience in trying to keep track of the numbers and to perceive a "wholeness" when solving problems. Therefore, it is essential that children who are on their way to grasping numerosity are given the opportunity to use their senses in various ways and experience numbers in a sensuous fashion.

One way of Handling afford more than one Meaning

As mentioned above, the main difference between the current study and earlier research within the field is that in this study children's various ways of handling numbers forms the basis for further analysis and interpretation of their experience regarding the meaning of numbers. Thereby, a distinct difference between how children go about solving the problems and how they experience and understand the meaning of numbers is revealed. It is shown that children discern, and are simultaneously aware of, different aspects of numbers and that there are different qualities both in children's ways of handling numbers and in their experience of numbers. There is no one to one correspondence between children's ways of handling numbers and their experience of meaning. On the contrary, *children's different ways of handling numbers can afford the same experience* and they can reach the same understanding from different directions. This is most clearly illustrated when children handle numbers by "counting".

When children are "counting", there are a lot of different ways of handling that bear the same experience of numbers (Table 4). For example, Daniel (p 55) is "counting and listening", while Tobias (p 61) is "counting" and "using his fingers looking". Regardless of the different ways of handling numbers, both children are focally aware of the same aspects of numbers, namely as "Positions in Sequence" and as "Composite Units". When solving one of the problems Christian is "counting and looking"(p 54), but when solving another problem he is "counting" and "using fingers; looking". Both of his ways of handling numbers offer the same meaning of numbers and he experiences numbers as "Positions in Sequence" and as "Composite Units".

The main conclusions drawn from these findings are that numbers appear in the same way for children when using different ways of handling numbers and it is obvious that there is not just one pathway to numbers but many, and when learning to grasp numerosity children should be given the opportunity to experience numbers in many various ways.

Chapter 4

THE DYNAMICS OF EXPERIENCING NUMBERS

The aim of this study is to illustrate some of the aspects of how children grasp numbers and the way in which these various concepts are related to one another. The primary result shows that there are distinctly different qualities in children's experience and that children are simultaneously aware of various aspects of numbers. In the following section we will deal both with *children's experience of numbers on a collective level and in individual cases.*

Looking at Number Development through the Experience of the Individual Child

Experience can be applied to an individual case in the sense that a description of a single child's experience can be studied in relation to the description of the experience on a collective level. As shown, a child could use several different ways of handling numbers when trying to solve various problems. The most common is to use two or three different ways of handling numbers. However, as well as handling numbers in different ways when solving different problems, a child can also change from one way to another while solving one and the same problem. One example is Christian, who starts by "estimating" equal numbers, but after a while changes focus and begins to make exact quantification (p. 49).

As this is not a longitudinal study the development of numerosity has to be analysed by setting out from a comparison between individuals. Some children, such as Andreas and Kristin, do not arrive at a single correct answer to any of the problems. They are "saying numbers", or on some occasions, "estimating". Dan and Sofia solve nine of the given problems, both making use of different ways of handling numbers, but mostly using "finger counting". Per arrives at a correct solution in 13 problems mainly by using "hearing". He says that he counts on his fingers and is able to show how he goes about it, but when solving the problem however, he does not use his fingers. One child, Johan, solves most of the given problems correctly, even in larger number ranges, mainly by using "known facts" and "structuring", but also by "hearing". He never "estimates", not even in very large number ranges, but simply says that he does not know. He says that he does not count on his fingers, but admits that he did so when he was "little". When comparing these children it seems very obvious that when children "say numbers" they do not display an understanding of the composite nature of numbers and that this stands in contrast to the children who use "structuring".

It seems that on their way towards grasping numerosity in a problem solving situation, children pass through different levels of understanding. Initially they experience numbers as "Number Words", they do not try to make exact quantification and direct their awareness towards numbers as the answer to a numerical problem and not towards the numerosity of the numbers. When the number range is large, even children who experience other aspects of numbers in smaller number ranges can focus on numbers as "Number Words". At the next level, children experience numbers as "Extents", having a vague sense of numerosity and the part-whole relationship of numbers in the problem, without using any procedure for exact quantification. Also, children who are aware of other aspects of numbers in smaller number ranges can focus on numbers as "Extents" when they do not know a procedure for solving the problem. At the next level, children experience numbers as "Positions in Sequence" or simultaneously experience them as "Positions in Sequence" and "Composite Units". They have an understanding of cardinality, and, to varying extents, the part-whole relationship of the numbers in the problem. At the next level, children focus on numbers as "Composite Units" and have a complete understanding of the part-whole relation and cardinality in the sense meant by Piaget (1967) and Fuson (1992).

It is evident that *children do not pass through these levels in a gradated, ordered and requisitioned manner.* On the contrary, their experiences are intertwined and children shift between the different ways of experiencing numbers, according to their interpretations of the content and structure of the problem, and even in solving one and the same problem can change between different ways of experiencing numbers. It has been mentioned earlier that it is not unusual for children to use "structuring" when solving a problem in a small number range, but then shift to "saying numbers" in a larger number range or when they do not understand the structure of a problem. This may correspond with how nine year old children, faced with problems regarded as difficult, use "finger counting", although when solving other problems do not use their fingers (Ahlberg, 1992). Siegler & Crowley (1991) stress the complexity in children's use of strategies.

> Cognition and cognitive growth are far more variable than our models typically suggest. People often use multiple strategies where they have depicted as using only one. The most accurate children tend to use the more advanced strategies, but sometimes do not (p. 618).

Children who have a developed number sense do not experience only one aspect of numbers, nor do they handle numbers in only one way, rather they are aware of many aspects and ways of handling numbers - *experiences overlap, interpenetrate, and complete one another.*

From Number Words to Composite Units

How then do the different ways of experiencing numbers evolve? Is it possible to make the growth of children's mathematical thinking visible? Are there embryos in the earlier phases of development that can later be found in future conceptions? In the following section, some earlier ways of handling and experiencing numbers are looked at, keeping in mind the ways that succeed them. By looking at the variation in the ways of handling and experiencing numbers on an individual level the roots of how children understand numbers in problem solving situations can be illuminated.

There were no instances identified in the empirical material where children formed a plan of action; rather, they simply tried to solve the problem. If they did not understand the content and structure of the problems or ran into difficulties, then they tried again or gave an answer by "saying a number". When "saying a number" the children know that a number word refers to a numerosity, but still they are not aware of the precise numerosity. In the problem solving situations their awareness was not directed towards the numerosity of numbers, even though they had an early numerical understanding, knowing for example that 2 is a small number and 9 is a big number and furthermore that 2 is less than 9. This initial vague sense of numerosity successively developed into knowledge of the approximate quantity of the numbers, as when the children are "estimating" in order to give a probable answer without quantification, but know that the answer is inexact and that there can be more or less.

When "saying numbers" children sometimes say "successive numbers", going one step upwards or downwards on the number sequence, and know that there must be some changes concerning the quantities mentioned in the problems, even though they do not have access to an accurate procedure with which to perform these operations. Hanna (p 41) tries to solve a subtraction problem, but she is not aware of the precise numerosity the number words refer to, and is not yet experienced with problems in a larger number range. In spite of this, she tries to count upwards on the number sequence and the answer turn out to be a "successive number".

When children use "successive numbers" they are not aware of numbers as "Positions in Sequence", but they have a growing insight that one may count on the number sequence in order to perform number operations. In this way, saying "successive numbers" can be the base for using the number sequence as a tool for doing operations.

Children do not use their fingers to perform number operations when "saying numbers ", although they often move their fingers and touch them when trying to come up with an answer. Some children start to count one of the parts in the problem on their fingers, but do not go continue performing an operation, and simply give the number of fingers they have counted as an answer. This way of trying to use fingers for solving problems can be the embryo that grows to the later use of finger counting. Finger counting is not something children are just suddenly able to do. They have to learn, and they learn by practicing long and hard. It happens quite often that they make mistakes and lose track of the counted numbers.

Even when children are "counting and tapping" it is obvious that they are trying to find a way to keep track of the counted numbers. Dan (p 42) tries to solve the problem by counting on the number sequence and tries to keep track of the counted numbers by tapping the table. However, as he doesn't have knowledge of the precise numerosity the number words refer to, his awareness is directed towards numbers as "Number Words". Therefore, he does not get any help from this way of handling numbers and arrives at an answer by saying a "successive number". Still, it is when trying to learn how to keep track of the numbers and perceive their numerosity, that he comes to learn how to perform number operations and to develop the understanding of numbers as "Positions in Sequence" and "Composite Units"

"Double counting" is a very rare way to handle numbers among the children in this study, where, contrary to other described ways of "counting" in which the children have a simultaneous experience of numbers as "Positions in Sequence" and as "Composite Units", "double counting" offers only the experience of numbers as "Positions in Sequence". However, children combine "double counting" with other ways of handling numbers such as "counting and listening". Daniel (p 52) uses "double counting" while solving four of the problems. However, when solving other problems he is "counting and listening" (p 55). As the only way of perceiving the numbers when "double counting" is through hearing, some children who use "double counting" probably abandon this and shift to "counting and listening" (in addition to using other ways of handling numbers). Still, nine year olds will sometimes use "double counting" (Ahlberg, 1992).

Some children use their fingers for "counting all". However, it is obvious that most children "counting all" in a larger number range use "counting and looking" when solving problems within smaller number ranges, thus perceiving the numerosity of one part by seeing. By using their fingers for "counting all" children handling numbers by "counting and looking" simultaneously experience several aspects of numbers by grouping them in different ways. In this way they gradually come to grasp the composite nature of numbers.

Simultaneously being aware of numbers as "Positions in Sequence and as "Composite Units" is the embryo that later grows to the ability to handle numbers by "structuring". When children are "counting looking at one part and the whole" or "looking at two parts" they can perceive that numbers can be composed and decomposed. From being used in conjunction with "counting", the handling of numbers by "structuring numbers seeing" may develop into a complete analogue representation of the part-whole relationship of numbers. However, it seems as though most children do not need to structure the complete part-whole relationship on their fingers by "seeing" the numerosity, but are able to grasp numerosity when grouping the numbers in various ways perceiving numbers as "Composite Units".

Chapter 5

CHILDREN LEARNING NUMEROSITY

How do children come to their understanding of numbers? This question was raised in the introduction of this report and hopefully the described study has shed light on some parts of this very complex process from the standpoint of an "experiential perspective" by elucidating the different ways children experience numbers, when solving elementary addition and subtraction problems.

Showing the way that children understand numbers when solving problems is a very complicated enterprise, as several factors influence the problem solving process. Children bring a whole range of experiences into the problem solving situation and a child's interpretation of a problem, to a large extent, is affected by the situation, the mathematical and linguistic structure and the content of the problem; it is influenced both by the "internal frame of reference" - the problem the child is trying to solve - and the "external frame of reference" - all the experiences appresented for the child in the problem solving situation. Children gain knowledge about the world directly through their senses and through acting and handling objects in the world around them. That is to say, their awareness of different aspects of a phenomena constitutes their experience and they get an appresentation of the whole from some of its parts (Ahlberg, 1992). It is not possible to show children's understanding of "the whole phenomena", but what is conceivable to show are children's somewhat differing ways of understanding numbers in this particular problem solving context.

The Results in relation to Different Views on Learning Numerosity

In trying to make a summery of the findings in the conducted study in the following the findings are going to be discussed and scrutinised in the light of the different views on the development of number concepts described in the introduction chapter.

Subitizing - Immediate Perception of Small Numbers

One issue of interest in research concerning children and numbers is whether there are counting principles or some innate number knopwledge guiding the development of numerosity. The current study has not shed light on this issue, however, another question has been clearly defined and that is whether subitizing is of importance for the development of numerosity (Glasersfeld 1987; Fisher, 1992: Neuman, 1987).

Many of the children in this study solved the problems using "counting and listening", i.e. *hearing* how many numbers they say, using the hearing as a component in the quantification process. In handling numbers this way one boy is able to perceive up to six and seven items (p 56). Children also subitize when using some of the described ways of "finger counting" in that they group a set of fingers and perceive them as a unit. By touching or looking at their fingers they can get an immediate perception of numerosity. Without counting unit by unit, they get a sensuous experience of the set of fingers in that they are *seeing* or *feeling* the numerosity.

Neuman (1987) stresses the significance of subitizing for developing arithmetic skills and puts forward the idea that when children are creating "finger numbers" they expand the subitizing range. However, it can not be said for sure, that the children in this study are extending the subitizing range, even when they perceive five or six items *seeing* or *hearing* the numerosity, because they may group the numbers by experiencing them in parts. Still, for children experiencing numbers by *seeing* or *hearing*, subitizing seems to be significant for the grasping of numerosity.

Counting and Reflective Abstraction

Another question which has been illuminated in the study concerns the importance of counting to grasp numerosity. von Glasersfeld (1994) argues that the notion of an abstract concept of numbers may arise from the activity of counting.

> Steffe, Richards, Cobb and I have presented a model on how an abstract concept of "Number" may be derived from the activity of counting (Steffe et al 1983). So far, I have not seen a more plausible model. But irrespective of that particular model, I have no doubt that both the notions of ordinal and cardinal number derive from the operations an active subject carries out and not from any specific sensory material (p. 6).

He outlines that there are fundamental operations of the mind from which one conceptualises unity and plurality, the former derived from the conceptual construction of "objects" and the latter from an awareness of repetition of the recognition of objects. Number concepts are merely the result of sense impressions, but the result of *reflective abstractions* on those sense impressions. Thus he claims that mathematics is a matter of internal mental operations and the meaning of numbers is an association of mental operations with mathematical symbols.

Also in this study the findings show that "counting" seems to be an important feature in children's development of numerosity. However, in this study with the six year olds their different ways of understanding the meaning of numbers do not seem to originate from the abstraction of mathematical operations through internal mental operations, but to a great extent from *their unification of numbers and ways of handling*. When children handle numbers in different ways they experience the meaning of numbers through their different modes of expressions and numbers have a meaning to them even though they do not have associated specific mental operations with the symbols. The six years old participating in this study, often had a simultaneous, bodily anchored, sensuous, experience of numbers, which seems to be of great importance for the foundation of their development of numerosity. This is evidenced in the description of how children use "counting" and grasp the numerosity of numbers through hearing, feeling and seeing.

The Sequence of Whole Numbers

There are a huge number of models that suggest that children's counting on the number sequence is an important tool for developing number sense (Gelman & Gallistel, 1978; Fuson 1992; Steffe & Cobb, 1988).

The findings in this study show that counting on the number sequence is by far the most common way for children to handle numbers and that they make use of many different ways of counting. Sometimes they keep track of the numbers, by using a purely mental act as in "double counting". When handling numbers this way, they identify *-landmarks-* the start and the end of the counting, and come to an answer without gaining any experience of the part-whole relation of numbers. Children also keep track of the counted numbers by using their senses for "marking" the numbers to perceive the numerosity, as for example when they are handling numbers "counting and touching". Then they - in contrast to "double counting" - gain experience of the composite nature of the numbers in the problem.

Finger counting - Sequential and Analogical Representations

In some models the children's use of fingers making analogical representations of the numbers without counting, is heavily emphasised and looked upon as one way to develop numerosity (Brissaud, 1992; Marton & Booth, 1997; Neuman, 1987).

When children use "finger counting", they go about it in several different ways; however, a characterising feature is that *they do not distinguish between numbers and fingers.* They use their fingers as references - creating a unit between numbers and fingers - they are both things at once. Sometimes this unity between fingers and numbers is in a one to one correspondence, but a group of fingers may also represent a number. When using "finger counting" children are simultaneously aware of numbers both as "Positions in Sequence" and as "Composite Units".

Neuman (1987) suggests that some ways of counting, namely "renaming" and "counting all/ on using two systems", does not contribute to the development of arithmetic skills. On the contrary it is looked upon as a "blind alley" and instead of gaining an understanding of cardinality the children get into mathematical difficulties (see p 18). In this study as well, some children use these ways of handling numbers. However, the interpretation here is not that these children are predestined to have mathematical difficulties. This way of handling numbers is looked upon more as a phase of development that the children may well pass through. Even when using these ways of handling numbers they group their fingers whilst simultaneously experiencing the numbers through their senses, and successively they can come to grasp the composite nature of numbers. When a child at the age of ten or eleven does not use any other way of handling numbers than for example in the terminology of Neuman "counting using two sequences", this is an indication of mathematical difficulties. However, as this is not a longitudinal study it can not clarified, whether the root of this difficulties is the use of a particular way of handling numbers at the age of six or seven.

The findings in this study seem to show that the nature of experiencing numerosity seems to be characterised by the aspect of numbers that children can discern and simultaneously handle and not whether they make use of sequential or analogical representations, because it seems as though synthesising and analysing are pre-requisite to one another.

Problem Solving and the Development of Number Concepts

In the research described in the introductory chapter, some models were described of the strategies children use when solving different types of elementary arithmetic problems. (Carpenter & Moser, 1982; Fuson; 1992) The "strategies" or ways of handling numbers revealed in this study do not differ in any remarkable way from earlier studies in that they use the number sequence counting upwards and downwards when solving simple addition and subtraction problems. Children are counting all - counting on from the first- counting on from largest. Children also use their knowledge of number facts. However in addition to these earlier descriptions an analyses has been made of the different ways in which children handles numbers when not having access to any procedure to solve the problems as when "saying numbers words" and "estimate".

The use of different ways of handling numbers has been carefully analysed and it is shown that it is very common that children shift between different ways of handling numbers according to the content and structure of the problem and that they might solve problems using a procedure within a smaller, well known number range, but do not make use of any procedure but simply give an answer "saying numbers" when trying to solve problems in larger number ranges.

The main difference between this study and earlier ones is however the distinction made between children's ways of handling numbers and the way in which they simultaneously experience different aspects of numbers in their interpretation of meaning and furthermore it has been revealed that children may create the same meaning through different ways of handling numbers.

Learning Numbers in an Experiential Perspective

The description in this study is outlined in terms of ways of handling and experiencing numbers and the findings show that children experience four different aspects of numbers; Number Words - Extents - Positions in Sequence - Composite units. This result may seem to be rather similar to the description presented by Marton and Booth (1997, p 65), who take their point of departure in Neuman's work (1987; 1989) when outlining how children experience numbers as names - extents - counted numbers - finger numbers - number facts (earlier described on p 18).

When comparing the two descriptions, the views of what it means to experience simultaneously different aspects of numbers are much the same in the two descriptions in the sense that the focus is towards the different aspects of numbers children are focally aware of in relation to the mathematical context. However, there are some discrepancies that might be discussed. First of all the way in which the analyses have been carried out differs. In this study the aim is to describe children's awareness of numbers *by discriminating between their ways of handling and experiencing,* Marton and Booth do not make this distinction and thus the descriptions of *what aspects of numbers children simultaneously experience* turn out to be quite different. Marton and Booth focus on the mathematical aspects of ordinality and cardinality.

> What we are saying is that the development of arithmetic skills demands that exactly these apparently abstract aspects can - and in fact must - be experienced sensuously and simultaneously (p. 66).

In the current study the focus is on describing the ways in which children experience numbers when handling them in various ways and therefore it becomes possible to show the children's interpretation of the meaning of numbers expressed in terms other than the abstract mathematical notions of "ordinality" and "cardinality". That is in terms of their own interpretations of meaning.

As the basis for the two analyses differs it is not a very fruitful enterprise to scrutinise the descriptions of children's experience in detail. However, it may be of interest to mention that in this study the interpretation of children's different ways of "keeping track" of numbers when solving verbal problems does not deliberately separate their simultaneous experience of different aspects of numbers. Furthermore the experience of numbers as "names" is not found in this study. The experience of numbers as "Number Words" outlined in this study is not compatible to "Names", as when they are aware of numbers as "Number Words" children know that the word refers to a numerosity even though the precise numerosity is not clear. Also, it is doubtful whether the experience of "finger numbers" is a general level for the development of numerosity in the way in which Marton and Booth suggest (Fig. 4.2). Besides the findings in this study, a further indication of this is that blind children do not use their fingers when developing numerosity (Ahlberg & Csocsán, 1994) and nor do hearing impaired children frequently make use of "finger numbers" when doing arithmetic (Frostad & Ahlberg, 1996).

The findings of the current study however do confirm Marton & Booths suggestion that children develop their arithmetical skills by using their sensuous experience, and it is clearly shown that children in their various ways of handling number simultaneously experience different aspects of numbers.

Learning Numerosity in Educational Settings

One probable reason for the great interest in the development of numerosity is that children's understanding of the early number concepts is of essential importance to their future learning of mathematics, and the belief that if we know something about the road that children take to understanding numbers and their grasping of numerosity, this knowledge could contribute to the field of teaching. This is a pedagogical question, and in the light of this study it has to be formulated in terms of how it is possible to bring about different ways of experiencing numbers.

Obviously the children in this study do not learn numbers through mechanical repetition. On the contrary, their own ways of exploring numbers and their sensuous experience of numbers is fundamental to learning and can not be dismissed. Taking the conclusions of this study as the point of departure, it is obvious that six year old children do not benefit from narrow, early formal mechanical training of mathematics in textbooks, as this does not open them to experiencing numbers and making use of these experiences. Baroody (1987) points out that instruction which is not geared to children can distort their views of mathematics, mathematical learning, and self. Several researchers emphasise that a very early introduction of symbols and formal procedures does not aid the development of arithmetic skills (Ahlberg, 1995; Neuman, 1987).

Bringing about Changes in Children's Ways of Experiencing Numbers

A flexible mind is often said to be a pathway to understanding different learning contents. What characterises children with mathematical difficulties is that they lack this flexibility. They are very rigid in their way of using different ways of handling numbers when solving arithmetic problems and do not use a variety of strategies applied to the content and structure of the problem. Carpenter & Moser (1984) and Siegler (Siegler & Jenkins, 1991; Siegler & Crowley, 1991), among others, emphasise the importance for children of learning to use different strategies when solving verbal problems. The findings in this study also show that children who use various ways of handling numbers when trying to solve the different tasks are, to a great extent, the ones who manage to arrive at a correct answer (p. 77).

However, *to be conscious of different strategies or ways of solving a problem one must be aware of the fact that the variation exists.* Is it possible to expand children's understanding by making it possible for them to be aware of different ways of handling numbers and different ways of experiencing numbers? The findings in this study seem to show that children should group numbers in various ways in order to come to experience them as composite units. To make children aware of different aspects of numbers they should be given opportunities to handle numbers in different ways - ways in which *one aspect of the phenomena is kept constant and other aspects are varied.* For example in grouping 5, the "number five" - the whole- remains constant but the "situations of five" - the parts- such as 2 and 3 or 4 and 1 can be varied.

Bringing Children's Experienced World into the Teaching Process

According to Marton and Booth (1997, p. 161) successful teaching brings about certain qualities in people's learning experience, and must address certain features of this - a structure of relevance and a pattern of dimensions of variation. They outline these two principles of teaching in this respect. The first in *building the relevance structure.* where they explain that:

> ... any learning situation or any situation at all, has a structure of relevance for those who experience it - aspects of the situation that indicate what it is aimed at, what it demands and where it will lead (p. 180).

The second principle of teaching is the design of the learning situation around the *architecture of variation.* This is described by saying:

> Whatever the content of instruction is, the object of study which the teacher forms or moulds has several faces for the learner to glimpse and mesh into the whole. ... There has to be a dimension of variation, which is to say that the variation has to be applied to something that might otherwise be fixed, taken for granted (p. 185).

Marton and Booth declare that, in their view, one main task of pedagogy is to identify educationally critical features inherent in capabilities to be exercised in certain types of situations. The task of pedagogy is to identify such critical features and to focus on them - by building a relevance structure, by making use of the "architecture of variation".

In building the relevance structure children's intentions, earlier experiences and understanding are very important features in the development of numerosity, and so are the activities and settings in which the practice of counting is experienced. Talking and interaction with other people is assumed to be a very important criterion for children's learning, and this is also the case when it comes to learning mathematical concepts and procedures (Vygotsky 1978). Also in the teaching and learning of mathematics the importance of language must not be dismissed (Ahlberg, 1992; Johnsen-Høines, 1991; Ekeblad 1997).

Language and interaction with others must be in focus in an analysis of children's experience of numbers, and surely are central when it comes to implementing the findings of the study in teaching practice. In two pre-school studies conducted with six year olds the aim was to let the children encounter mathematics while taking into account *their own experienced world* (Ahlberg (1994; 1995). In the study of "Writing, drawing and counting mathematics" (Ahlberg, 1994), pre-school children worked with different types of word problems. In order to develop the children's flexibility of mind and get them more inclined to investigate and to be aware of and learn numerosity, a variation was introduced in the teaching process. This variation was along three different dimensions: the content of the problems, the means of expression and in exposing the variation of the children's ideas and thoughts.

In the second study "Mathematics in thematic activity" (Ahlberg, 1995) the teacher worked in co-operation with the researcher to develop and improve teaching methods. The theme chosen was "The Tree". In this study as well as in the one earlier described *the children's own experienced world* was used as a foundation for the teaching. The aim of the instruction was to give the children the opportunity to learn mathematical concepts in their encounters with numbers in many different contexts, by using different senses, and different modes of expression.

Children should learn numerosity not only by counting one by one but also by perceiving the numerosity by grouping different kinds of objects as leaves, fruits connected to trees things they had a personal relationship as they had collected themselves. In grouping the objects one aspect of the phenomena was kept constant and the other aspects were varied. The "situations of five" - the parts- were varied, but the "number five" - the whole- remained constant. For example children were asked to group five fruits in different ways and in taking part in one another different ways of handle and group the "number five" in parts, as 1 and 4 or 2 and 3 they were given opportunities to learn how to operate on numbers and develop their understanding of numbers discussing and learning from each other.

In many different learning situations concerning different mathematical content the children explored numbers as composite units while grouping objects in various settings and every day situations. They used various means of expression in talking, drawing and counting. *Differences between the children were not seen as a obstacle to instruction - on the contrary they were seen as opportunities* - where the children were given the opportunity to learn from one another. When making use of different ways of handling numbers and talking about their different ways of experiencing numbers, they were given the possibility to discern new aspects of numbers on their way to developing numerosity.

The two described studies illustrated that this way of working with young children and mathematics contributes greatly to their early mathematical learning. In both studies the results of a comparison between an equivalent control group and the participating children revealed that this way of working with children and mathematics had contributed to the children's development of numerosity, not only in giving more correct answers than children in equivalent control groups, but also in that they had to a greater extent used different strategies or ways of handling numbers and experienced numbers in a more advanced way.

Concluding Remarks

A main finding of this study is that numbers may appear for children in the same way when using different ways of handling numbers while solving a problem. There is no one to one correspondence between children's different ways of handling numbers and their experience of meaning. They may experience the same aspects of numbers, though they use different ways of handling the numbers. The results of this study suggest that the focus of the teaching process should be on how to provide opportunities for children to experience new aspects of numbers, by bringing their experiences into the situation and making use of them when using their sensuous experience. Children should be given opportunities to bring available resources to new circumstances when encountering a variation of mathematical content and ways of handling numbers in order to be aware of the different aspects of numbers when solving problems.

The aim of the study was not to provide a final answer to the question of how children come to understand numbers, but to elucidate some of the aspects concerning children's grasp of numerosity, within a certain theoretical framework - the phenomenograpic research approach - and to illuminate some aspects of children's understanding of numerosity in an experiential perspective. When a certain theoretical standpoint forms the basis for analyses and interpretations research quite often results in *one* answer to a posed question, which in this context of children and numbers would be to outline *one way* to develop numerosity. In this study however, the outcome is quite the opposite and the results describe a tremendous variation. When trying to grasp numerosity children handle numbers in a vast array of ways, and thereby experience different aspects of numbers. However, in spite of using different ways of handling numbers, the numbers may appear in the same way to them and they may experience the same meaning. Consequently, there is not only one pathway, but many pathways to numbers. According to this the implications for the teaching process drawn from this study are that in order to develop children's mathematical thinking they should be given the opportunity to interact with others and the world around them by talking and reflecting over various mathematical contents. The focus of instruction must be on both grasping every day situations in which a mathematical content can be visualised and reflected over, and on organising pre-planned learning situations in which children are given the possibility to meet numbers in various contexts and in meaningful problem solving situations in which they experience and discover different aspects of mathematics and numbers. One conclusion drawn from the findings in this study is that many children use their sensuous experience in trying to keep track of the numbers and to perceive a "wholeness" when solving problems. Therefore, it is essential that children who are on their way to grasping numerosity are given the opportunity to use their senses in various ways and experience numbers in a sensuous fashion.

Understanding numbers and learning arithmetic skills is not only a question of a the quantification of objects or fingers. Neither is it a matter of learning how to count on the number sequence or developing logical thinking. It is instead, a question of being able to explore and discern different aspects and possible qualities of the numbers - of experiencing numbers in the sense of sensuously and simultaneously perceiving different aspects of numbers.

References

Ahlberg, A. (1992). *Att möta matematiska problem. En belysning av barns lärande.* (Meeting mathematical problems. An illumination of childrens learning.) Göteborg: Acta Universitatis Gothoburgensis.

Ahlberg, A. (1994). *Barn och matematik.* (Children and mathematics.) Lund: Studentlitteratur.

Ahlberg, A. (1995). *Att möta matematiken i förskolan, Rita, tala och räkna matematik.*(Meeting mathematics in pre-school. Drawing, talking and calculating mathematics.) Rapport 1994:12. Institutionen för pedagogik. Göteborgs universitet.

Ahlberg, A. (1995). *Att möta matematiken i förskolan, Matematiken i temaarbetet.* (Meeting mathematics in pre-school. Mathematics in the thematic activity.) Rapport 1995:14. Institutionen för pedagogik. Göteborgs universitet.

Ahlberg, A. & Csocsán, E. (1994). *Grasping numerosity among blind children.* Report, 1994:04. Institutionen för pedagogik. Göteborgs universitet.

Ahlberg, A. & Csocsán, E. *Blind children and their experience of numbers* Institutionen för specialpedagogik. Göteborgs universitet. (Forthcoming).

Ahlberg, A, & Hamberger, B. (1995). *Att möta matematiken i förskolan - 6-åringars förståelse av tal och räkning.* (Meeting mathematics in pre-school. Six year olds understanding of numbers and counting.) Rapport 1995:08. Institutionen för pedagogik. Göteborgs universitet.

Antell, S. E. & Keating, D. P. (1983). Perception of numerical invariance in neonates. *Child Development,* 54: 695-701.

Baroody, A. J. (1984). The case of Felicia: A Young Child's Strategies for Reducing Memory Demands During Mental Addition. *Cognition and Instruction,* 1:109-116.

Baroody, A. J. (1984). Children's difficulties in subtraction: some causes and questions *Journal for Research in Mathematics Education,* Vol. 15, 3: 203-213.

Baroody, A. J. (1987). *Children's Mathematical Thinking.* New York: Teachers College Press.

Baroody, A. J. (1992). The Development of Preschoolers' Counting Skills and Principles. In Bideaud, J. Meljac, C. & Fisher, J-P. (Eds.), *Pathways to number. Children's developing numerical abilities.* Hillsdale New Jersey: Lawrence Erlbaum.

Bremner, J. G. (1992). *Infancy.* Oxford: Page Bros Ltd.

Brissaud, R. (1992). A toll for number construction: Finger Symbol Sets. In Bideaud, J. Meljac, C. & Fisher, J-P. (Eds.), *Pathways to number. Children's developing numerical abilities.* Hillsdale New Jersey: Lawrence Erlbaum.

Brown, J. S., Collins, A. & Duguid, P. (1989). Situated cognition and the culture of learning. *Educational Researcher,* 18 (1): 32-42.

Bryant, P. (1994) Children and arithmetic. *Journal of Child Psychology and Psychiatry,* 36: 3-32.

Carpenter, T., Moser, J. & T. Romberg, (Eds.), (1982). *Addition and subtraction; A cognitive perspective.* Hillsdale, NJ: Lawrence Erlbaum.

Carpenter, T. & Moser, J. (1982). The Deveolpment of Addition and Subtraction Problem Solving Skills. In *Additon and subtraction; A cognitive perspective.* Carpenter, T., Moser, J. & T. Romberg, (Eds.), Hillsdale, NJ: Lawrence Erlbaum.

Carpenter, T. & Moser, J. (1984). The aquisitions of addition and subtraction concepts in grades one through three. *Journal for Research in Mathematics Education, 15.* No 3,179-202.

Carraher, T., Carraher, D. & Shliemann, A. (1985). Mathematics in the streets and in Schools. *British Journal of Developmental Psychology, 3,* 21 -29.

Deloche, G. & Seron. X. (1987). *Mathematical Disabilities.* A cognitive Neuropshycolocial Perspective. London: Lawrence Erlbaum.

Droz, R. (1992). The multiple roots of natural numbers and their multiple interpretations. In Bideaud, J. Meljac, C. & Fisher, J-P (Eds.), *Pathways to number. Children's developing numerical abilities.* Hillsdale New Jersey: Lawrence Erlbaum.

Ekeblad, E. (1996). *Children learning numbers.* Göteborg: Acta Universitatis Gothoburgensis.

Ernest, P. (1991). *The philosophy of Mathematics Education.* London: The Falmer Press: London.

Ernest, P. (1994). (Ed). *Constructing Mathematical Knowledge: Epestimology and Mathematical Education..* London: The Falmer Press.

Fischer, J. P. (1992). Subitizing: The Discontinouity After Three. In Bideaud, J. Meljac, C. & Fisher, J-P (Eds.), *Pathways to number Children's developing numerical abilities.* Hillsdale New Jersey: Lawrence Erlbaum.

Frostad, P. & Ahlberg, A. (1996). Conceptions of Numbers - The perspectives of Hearing Impaired Norwegian Schoolchildren. *Det Konglige Norske Videnskapers Selskap.* Skrifter 2/1996.

Fuson, K. (1988). *Children's counting and concepts of number.* New York: Springler Verlag.

Fuson, K. (1992). Relationships Between Counting and Cardinality From Age 2 to Age 8. In Bideaud, J. Meljac, C. & Fisher, J-P (Eds.), *Pathways to number. Children's developing numerical abilities.* Hillsdale New Jersey: Lawrence Erlbaum.

Fuson, K.C. (1992). Research on whole number addition and subtraction. In Grouws, D.A. (Ed.) *Handbook of research on mathematics teaching and learning.* New York: Macmillan Publishing Company, p 243-276.

Fuson, K. C, & Hall, J. W. (1983). The acquisition of early number word meanings: A conceptual analysis and review. In H.P. Ginsburg (Ed.), The *development of matehematical thinking* (p 49-107). New York: Academic Press.

Gallistel, C. R. & Gelman, R. (1992). Preverbal and verbal counting and computation. *Cognition, 44:*43-47.

Gelman, R. & Baillargeon, R. (1983). A review of some Piagetan concepts. In P.H. Mussen (Ed) *Handbook of child psychology, Vol 3*, Cognitive development. New York: Wiley.

Gelman, R. & Gallistel, C. R. (1986). *The child's understanding of number.* Cambridge, MA: Harvard University Press.

Gelman, R. & Gallistel, C. R. (1983). The child's understanding of number. In M. Donaldsson, R. Grieve. & C. Pratt (Eds.), *Early childhood development and education.* Oxford: Basil Blackwell.

Gelman, R. & Gallistel, C. R. (1992). Preverbal and verbal counting and computation. *Cognition,* 44: 43-74.

Gelman, R. and Meck, E. (1986). The notions of principle: The case of counting. In J, Hiebert (Ed.) *Conceptual and procedural knowledge: The case of mathematics p* 29-57. Hillsdale, NJ: Lawrence Erlabaum Associates.

Gelman and Meck, B. (1992). Early Principles Aid Initial but Not Later conceptions of Number. In Bideaud, J. Meljac, C. & Fisher, J-P (Eds.), *Pathways to number Children's developing numerical abilities.* Hillsdale New Jersey: Lawrence Erlbaum.

Ginsburg, H. (1977). *Children's arithmetic.* New York: Van Nostrand.

Ginsburg, H. (Ed). (1983). *The development of mathematical thinking.* New York: Academic Press.

Gray, E. & Tall, D. (1994). Duality, Ambiguity, and Flexibility: A "Proceptual" view of Simple Arithmetic. *Journal for Research in Mathematics Education, V 25,* 2:116-140.

Hiebert, J, & Lefevre, P. (1986). Conceptual and procedural knowledge in mathematics: An introductory analysis. In J. Hiebert (Ed.), *Conceptual and procedural knowledge.* Hillsdale, NJ: Lawrence Erlbaum.

Hughes, M. (1986). *Children and number. Difficulties in Learning Mathematics.* Oxford: Basil Blackwell.

Johansson, B., Marton. F. & Svensson, L.(1985). An approach to describing learning as change between qualitative different conceptions. In L. H. T. West & A. C: Pines (Ed.), *Cognitive structure and conceptual change.* New York: Academic Press.

Johnsen-Høines, M. (1990). *Matematik som språk.* (Mathematics as language.) Stockholm: Utbildningsförlaget.

Klahr, D. & Wallace, J. G. (1973). The role of quantification operators in the developmemt of conservation of quantity. *Cognitive Psychology,* 4: 301- 327.

Kvale, S. (1996). *InterViews. An introduction to qualitative research interviewing.* Thousand Oaks: Sage Publications.

Lave, J. (1988). *Cognition in practice.* Cambridge: Cambridge University Press.

Lave, J. (1992). Word problems: a microcosm of theories of learning. In P. Light and G. Butterworth (Eds.), *Context and cognition. Ways of learning and knowing.* Hillsdale, NJ: Lawrence Erlbaum.

Lester, F. (1988). Reflections about mathematical problem solving research. In R. Charles & E. Silver (Eds.) *The teaching and assessing of mathematical problem solving.* Reston VA: National Council of Teachers of Mathematics. Lawrence Erlbaum.

Marton, F. (1981). Phenomenography - Describing conceptions of the world around us. *Instructional Science,* 10: 177-200.

Marton, F. (1994). On the structure of awareness. In Bowden, J. A. & Walsh, E. (Eds.) *Phenomenographic research*: Variations in Method. The Warburton Symposium.

Marton, F. (1988). Phenomenography and "the art of teching all things to all men". *Fenomenografiska notiser 8.* Institutionen för pedagogik Göteborgs universitet.

Marton, F. & Booth, S. (1997). *Learning and Awareness.* Mahwah, NJ: Lawrence Erlbaum.

Marton, F. & Neuman, D. (1989). Constructivism and Constitutionalism. Some Implications for Elementary Mathematics Education. *Scandinavian Journal of Educational Research, Vol* 33, 1:35-46.

Mellin Ohlsen, S. (1987). *The Politics of Mathematics Education.* Dordrecht: Reidel.

Mishler, E. G. (1986). *Research interviewing - Context and narrative.* Cambridge, MA: Harvard Unvversity Press.

Neuman, D. (1987). *The origin of arithmetic skills. A phenomenographic approach.* Göteborg: Acta Universitatis Gothoburgensis.

Neuman, D. (1989). *Räknefärdighetens rötter.* (The roots of arithmetic skills.) Stockholm: Utbildningsförlaget.

Piaget, J. (1969). *The child's conception of number.* London: Routledge & Kegan Paul.

Pimm, D. (1987). *Speaking Mathematically. Communication in Mathematics Classrooms.* London: Routledge.

Resnick, L. (1987). Learning in school and out. *Educational Researcher,* 16:13-20.

Resnick, L. (1989). Developing mathematical knowledge. *American Psychologist, 44* , 2:162-169.

Resnick, L. (1994). From protoquantities to Operators: Building Mathematical Competence on a Foundation of Everyday Knowledge. In G. Leinhardt, R. Putnam, & R. Hattup (Eds.), *Analyses of arithmetic for mathematics teachers.* Hillsdale, New Jearsey: Lawrence Erlbaum.

Romberg, T & Carpenter, T (1986). Research on teaching and learning mathematics: two disciplined of scientific inquiry. In *Handbook of Research on Teaching*, p 318-28. New York; Macmillan.

Schoenfelt (1987). *Mathematical problem solving*. New York: Academic Press.

Sigler, R. S. & Jenkins, E. (1989). *How children discover new strategies.* Hillsdale, NJ: Erlbaum.

Sigler, R. S. & Crowley (1991). The Microgenetic Method. A direct method for Studying Cognitive Development. *American Psychologist,* 46: 6. 606-620.

Steffe, R. & Cobb, P. (1988). *Construction of aritmetical meanings and strategies.* New York: Springer.

Steffe, L., von Glasersfeld, E., Richards, J. & Cobb, P. (1983). *Children´s counting types: Philosophy, theory and application.* New York: Praeger Scientific.

Steffe, L. P. (1992). Learning stages in the construction of the number sequences. In Bideaud, J. Meljac, C.& Fisher, J-P (Eds.), *Pathways to number. Children developing numerical abilities.* Hillsdale New Jersey: Lawrence Erlbaum.

Steffe, L. P., Von Glasersfeld, E., Richards, J., & Cobb, P. (1983). *Children's counting types: Philosophy, theory and application.* New York: Praeger Scientific.

Steffe, L. P. & Cobb, P. (1988). *Construction of arithmetical meanings and strategies.* New York: Springer Verlag.

Säljö, R. (1988). Learning in educational settings: Methods of enquiry. In P. Ramsden (Ed.), *Improving learning: New perspectives.* London: Kogan Page.

Säljö, R. & Wyndham, J. (1987). The formal Setting as Context for Cognitive activities. An Empirical study of Arithmetic operations under Conflicting Premises for Communication. *European Journal of Psychology of Education*, *Vol. 11,* 3: 233-245

von Glasersfeld, E. (1981). An Attentional Model for the Conceptual Construction of Units and Number. *Journal for Research in Mathematics Education.* 12:83-94.

von Glasersfeld, E. (1983). An Attentional Model for the Conceptual Construction of Units and Number. Appendix in L. P. Steffe, E. Von Glasersfeld, J. Richards and P. Cobb. *Children's counting types: Philosophy, theory and application.* New York: Praeger Scientific.

von Glasersfeld, E. (1993). Reflections on Number and Counting. In S. T. Boysen and E. J. Capaldi (Eds.), *The development of Numerical Competence. Animal and Human Models.* Hillsdale, NJ: Lawrence Erlbaum Associates.

von Glasersfeld, E. (1994). A Radical Constructivist View of Basic Mathematical Conceps. In Paul Ernest (Ed.),*Constructing Mathematical Knowledge: Epestimology and Mathematical Education..* London: The Falmer Press.

Wynn, K. (1992a). Addition and subtraction by human infants. *Nature*, 358:749- 750.

Wynn, K. (1992b). Children´s acquisitions of the number words and the counting system. *Cognitive Psychology,* 24:220-251.

Vygotsky, L. S. (1978). *Mind in society.* Cambridge, Ma Harward university Press.

Appendix 1

Every day problems

A.
1. If you have 2 apples and you get 3 more. How many apples do you have then? (2+3=_)

2. If you have 4 pennies and lose 2 of them. How many pennies have you got left? (4-2=_)

3. Imagine that you have 5 pennies in your purse when you go out. When you come home you have only 2 pennies in your purse. How many pennies have you lost? (5-_=2)

B.
4. If you have 2 crayons and I give you 7 more crayons. How many crayons will you have then? (2+7=_)

5. You have 10 candies and you give away 7 of them. How many candies have you got left? (10-7=_)

6. You have 3 pennies and you want to buy an ice-cream that costs 7 pennies. How many more pennies do you need? (3+_=7)

C.
7. You have 13 pennies and I give you 5 more pennies. How many pennies do you have then? (13+5=_)

8. You have 15 plums and you give away 7 of them. How many plums have you got left? (15-7=_)

9. You have 12 pennies and you want to buy a newspaper that costs 19 pennies. How many more pennies do you need? (12+_=19)

D.
10. You have 16 books and you get 8 more books. How many books do you have then? (16+8=_)

11. You have 25 pennies and you loose 7 of them. How many pennies have you got left? (25-7=_)

12. You have 23 pennies and you want to buy a book that costs 49 pennies. How many more pennies do you need? (23+_=49)

Appendix 2

Contextuel problems

The interviewer told a story about a baker who put some buns in the oven and then went for a walk. A bunthief come sneaking around. When the baker cames back some buns are missing. The children were shown a plate with 8 paper buns. They touch and count the buns and the interviewer tell the following story. "There are 8 buns on the plate. When the baker comes back there are only 2 buns. How many buns has the thief taken?"(8-_=2). The children solved the problem and thereafter the interviewer took the picture sequence away and posed the following problems.

1. The baker has baked and there are 6 buns on the plate. He took them out of the oven to let them cool and went for a walk. When he came back he sees that there are 3 buns on the plate, how many buns has the thief taken? (6-_=3).
2. The baker has baked and there are buns on the plate. Then he went for a walk and the bun thief came and took 3 buns. When the baker came back there were 2 buns left on the plate. How many has the baker baked?
3. If the thief takes 2 away and there are 7 left, how many were baked? (_-2=7)
4. If the baker bakes 8 buns and finds only 2, how many buns were taken? (8-_=2)
5. If there are 9 buns and the baker finds only 5 of them, how many were taken? (9-_=5)
6. How many buns did the baker bake if the thief took 3 buns and there were 4 left? (_-3=4)
7. If the baker bakes 14 and there are 8 left how many buns were taken? (14-_=8)
8. If the thief takes 7 buns and the baker finds 10, how many buns had he baked? (_-7=10)
9. If the baker baked 24 and there are 6 left, how many has the thief taken? (24-_=6)
10. Now the baker bakes 43 buns. How many are taken if there are 19 left? (43-_=19)

GÖTEBORG STUDIES
IN EDUCATIONAL SCIENCES

ISSN 0436-1121

1. *Karl-Gustaf Stukát:* Lekskolans inverkan på barns utveckling. Sthlm 1966. Pp.148.
2. *Urban Dahllöf:* Skoldifferentiering och undervisningsförlopp. Sthlm 1967. Pp. 306.
3. *Erik Wallin:* Spelling. Factorial and experimental studies. Sthlm 1967. Pp.180.
4. *Bengt-Erik Andersson:* Studies in adolescent behaviour. Project Yg, Youth in Göteborg. Sthlm 1969. Pp. 400.
5. *Ference Marton:* Structural dynamics of learning. Sthlm 1970. Pp. 112.
6. *Allan Svensson:* Relative achievement. School performance in relation to intelligence, sex and home environment. Sthlm 1971. Pp. 176.
7. *Gunni Kärrby:* Child rearing and the development of moral structure. Sthlm 1971. Pp. 207.
8. *Ulf P. Lundgren:* Frame factors and the teaching process. A contribution to curriculum theory and theory on teaching. Sthlm 1972. Pp. 378.
9. *Lennart Levin:* Comparative studies in foreign-language teaching. Sthlm 1972. Pp. 258.
10. *Rodney Åsberg:* Primary education and national development. Sthlm 1973. Pp. 388.
11. *Björn Sandgren:* Kreativ utveckling. Sthlm 1974. Pp. 227.
12. *Christer Brusling:* Microteaching - A concept in development. Sthlm 1974. Pp. 196.
13. *Kjell Rubenson:* Rekrytering till vuxenutbildning. En studie av kortutbildade yngre män. Gbg 1975. Pp. 363.
14. *Roger Säljö:* Qualitative differences in learning as a function of the learner's conception of the task. Gbg 1975. Pp. 170.
15. *Lars Owe Dahlgren:* Qualitative differences in learning as a function of content-oriented guidance. Gbg 1975. Pp. 172.
16. *Marie Månsson:* Samarbete och samarbetsförmåga. En kritisk granskning. Lund 1975. Pp. 158.
17. *Jan-Eric Gustafsson:* Verbal and figural aptitudes in relation to instructional methods. Studies in aptitude - treatment interactions. Gbg 1976. Pp. 228.
18. *Mats Ekholm:* Social utveckling i skolan. Studier och diskussion. Gbg 1976. Pp. 198.
19. *Lennart Svensson:* Study skill and learning. Gbg 1976. Pp. 308.
20. *Björn Andersson:* Science teaching and the development of thinking. Gbg 1976. Pp. 180.
21. *Jan-Erik Perneman:* Medvetenhet genom utbildning. Gbg 1977. Pp. 300.

(cont.)

GÖTEBORG STUDIES
IN EDUCATIONAL SCIENCES

ISSN 0436-1121

(Continuation)

22. *Inga Wernersson:* Könsdifferentiering i grundskolan. Gbg 1977. Pp. 320.
23. *Bert Aggestedt and Ulla Tebelius:* Barns upplevelser av idrott. Gbg 1977. Pp. 440.
24. *Anders Fransson:* Att rädas prov och att vilja veta. Gbg 1978. Pp. 188.
25. *Roland Björkberg:* Föreställningar om arbete, utveckling och livsrytm. Gbg 1978. Pp. 252.
26. *Gunilla Svingby:* Läroplaner som styrmedel för svensk obligatorisk skola. Teoretisk analys och ett empiriskt bidrag. Gbg 1978. Pp. 269.
27. *Inga Andersson:* Tankestilar och hemmiljö. Gbg 1979. Pp. 288.
28. *Gunnar Stangvik:* Self-concept and school segregation. Gbg 1979. Pp. 528.
29. *Margareta Kristiansson:* Matematikkunskaper Lgr 62, Lgr 69. Gbg 1979. Pp. 160.
30. *Britt Johansson:* Kunskapsbehov i omvårdnadsarbete och kunskapskrav i vårdutbildning. Gbg 1979. Pp. 404.
31. *Göran Patriksson:* Socialisation och involvering i idrott. Gbg 1979. Pp. 236.
32. *Peter Gill:* Moral judgments of violence among Irish and Swedish adolescents. Gbg 1979. Pp. 213.
33. *Tage Ljungblad:* Förskola - grundskola i samverkan. Förutsättningar och hinder. Gbg 1980. Pp. 192.
34. *Berner Lindström:* Forms of representation, content and learning. Gbg 1980. Pp. 195.
35. *Claes-Göran Wenestam:* Qualitative differences in retention. Gbg 1980. Pp. 220.
36. *Britt Johansson:* Pedagogiska samtal i vårdutbildning. Innehåll och språkbruk. Gbg 1981. Pp. 194.
37. *Leif Lybeck:* Arkimedes i klassen. En ämnespedagogisk berättelse. Gbg 1981. Pp. 228.
38. *Biörn Hasselgren:* Ways of apprehending children at play. A study of pre-school student teachers' development. Gbg 1981. Pp. 107.
39. *Lennart Nilsson:* Yrkesutbildning i nutidshistoriskt perspektiv. Yrkesutbildningens utveckling från skråväsendets upphörande 1846 till 1980-talet samt tankar om framtida inriktning. Gbg 1981. Pp. 442.
40. *Gudrun Balke-Aurell:* Changes in ability as related to educational and occupational experience. Gbg 1982. Pp. 203.
41. *Roger Säljö:* Learning and understanding. A study of differences in constructing meaning from a text. Gbg 1982. Pp. 212.
42. *Ulla Marklund:* Droger och påverkan. Elevanalys som utgångspunkt för drogundervisning. Gbg 1983. Pp. 225.

(cont.)

GÖTEBORG STUDIES
IN EDUCATIONAL SCIENCES

ISSN 0436-1121

Editors:

Ingemar Emanuelsson, Jan-Eric Gustafsson and Ference Marton

(Continuation)

43. *Sven Setterlind:* Avslappningsträning i skolan. Forskningsöversikt och empiriska studier. Gbg 1983. Pp. 467.
44. *Egil Andersson and Maria Lawenius:* Lärares uppfattning av undervisning. Gbg 1983. Pp. 348.
45. *Jan Theman:* Uppfattningar av politisk makt. Gbg 1983. Pp. 493.
46. *Ingrid Pramling:* The child's conception of learning. Gbg 1983. Pp. 196.
47. *Per Olof Thång:* Vuxenlärarens förhållningssätt till deltagarerfarenheter. En studie inom AMU. Gbg 1984. Pp. 307.
48. *Inge Johansson:* Fritidspedagog på fritidshem. En yrkesgrupps syn på sitt arbete. Gbg 1984. Pp. 312.
49. *Gunilla Svanberg:* Medansvar i undervisning. Metoder för observation och kvalitativ analys. Gbg 1984. Pp. 194.
50. *Sven-Eric Reuterberg:* Studiemedel och rekrytering till högskolan. Gbg 1984. Pp. 191.
51. *Gösta Dahlgren and Lars-Erik Olsson:* Läsning i barnperspektiv. Gbg 1985. Pp. 272.
52. *Christina Kärrqvist:* Kunskapsutveckling genom experimentcentrerade dialoger i ellära. Gbg 1985. Pp. 288.
53. *Claes Alexandersson:* Stabilitet och förändring. En empirisk studie av förhållandet mellan skolkunskap och vardagsvetande. Gbg 1985. Pp. 247.
54. *Lillemor Jernqvist:* Speech regulation of motor acts as used by cerebral palsied children. Observational and experimental studies of a key feature of conductive education. Gbg 1985. Pp. 146.
55. *Solveig Hägglund:* Sex-typing and development in an ecological perspective. Gbg 1986. Pp. 267.
56. *Ingrid Carlgren:* Lokalt utvecklingsarbete. Gbg 1986. Pp. 299.
57. *Larsson, Alexandersson, Helmstad and Thång:* Arbetsupplevelse och utbildningssyn hos icke facklärda. Gbg 1986. Pp. 165.
58. *Elvi Walldal:* Studerande vid gymnasieskolans vårdlinje. Förväntad yrkesposition, rollpåverkan, självuppfattning. Gbg 1986. Pp. 291.
59. *Eie Ericsson:* Foreign language teaching from the point of view of certain student activities. Gbg 1986. Pp. 275.
60. *Jan Holmer:* Högre utbildning för lågutbildade i industrin. Gbg 1987. Pp. 358.
61. *Anders Hill and Tullie Rabe:* Psykiskt utvecklingsstörda i kommunal förskola. Gbg 1987. Pp. 112.
62. *Dagmar Neuman:* The origin of arithmetic skills. A phenomenographic approach. Gbg 1987. Pp. 351.

(cont.)

GÖTEBORG STUDIES
IN EDUCATIONAL SCIENCES

ISSN 0436-1121

Editors:

Ingemar Emanuelsson, Jan-Eric Gustafsson and Ference Marton

(Continuation)

63. *Tomas Kroksmark:* Fenomenografisk didaktik. Gbg 1987. Pp. 373.
64. *Rolf Lander:* Utvärderingsforskning - till vilken nytta? Gbg 1987. Pp. 280.
65. *Torgny Ottosson:* Map-reading and wayfinding. Gbg 1987. Pp. 150.
66. *Mac Murray:* Utbildningsexpansion, jämlikhet och avlänkning. Gbg 1988. Pp. 230.
67. *Alberto Nagle Cajes:* Studievalet ur den väljandes perspektiv. Gbg 1988. Pp. 181.
68. *Göran Lassbo:* Mamma - (Pappa) - barn. En utvecklingsekologisk studie av socialisation i olika familjetyper. Gbg 1988. Pp. 203.
69. *Lena Renström:* Conceptions of matter. A phenomenographic approach. Gbg 1988. Pp. 268.
70. *Ingrid Pramling:* Att lära barn lära. Gbg 1988. Pp. 115.
71. *Lars Fredholm:* Praktik som bärare av undervisnings innehåll och form. En förklaringsmodell för uppkomst av undervisningshandlingar inom en totalförsvarsorganisation. Gbg 1988. Pp. 364.
72. *Olof F. Lundquist:* Studiestöd för vuxna. Utveckling, utnyttjande, utfall. Gbg 1989. Pp. 280.
73. *Bo Dahlin:* Religionen, själen och livets mening. En fenomenografisk och existensfilosofisk studie av religionsundervisningens villkor. Gbg 1989. Pp. 359.
74. *Susanne Björkdahl Ordell:* Socialarbetare. Bakgrund, utbildning och yrkesliv. Gbg 1990. Pp. 240.
75. *Eva Björck-Åkesson:* Measuring Sensation Seeking. Gbg 1990. Pp. 255.
76. *Ulla-Britt Bladini:* Från hjälpskolelärare till förändringsagent. Svensk speciallärarutbildning 1921-1981 relaterad till specialundervisningens utveckling och förändringar i speciallärarens yrkesuppgifter. Gbg 1990. Pp. 400.
77. *Elisabet Öhrn:* Könsmönster i klassrumsinteraktion. En observations- och intervjustudie av högstadieelevers lärarkontakter. Gbg 1991. Pp. 211, XXI.
78. *Tomas Kroksmark:* Pedagogikens vägar till dess första svenska professur. Gbg 1991. Pp. 285.
79. *Elvi Walldal:* Problembaserad inlärning. Utvärdering av påbyggnadslinjen Utbildning i öppen hälso- och sjukvård. Gbg 1991. Pp. 130.
80. *Ulla Axner:* Visuella perceptionssvårigheter i skolperspektiv. En longitudinell studie. Gbg 1991. Pp. 293.
81. *Birgitta Kullberg:* Learning to learn to read. Gbg 1991. Pp. 352.
82. *Claes Annerstedt:* Idrottslärarna och idrottsämnet. Utveckling, mål, kompetens - ett didaktiskt perspektiv. Gbg 1991. Pp. 286.

(cont.)

GÖTEBORG STUDIES
IN EDUCATIONAL SCIENCES

ISSN 0436-1121

Editors:

Ingemar Emanuelsson, Jan-Eric Gustafsson and Ference Marton

(Continuation)

83. *Ewa Pilhammar Andersson:* Det är vi som är dom. Sjuksköterskestuderandes föreställningar och perspektiv under utbildningstiden. Gbg 1991. Pp. 313.

84. *Elsa Nordin:* Kunskaper och uppfattningar om maten och dess funktioner i kroppen. Kombinerad enkät- och intervjustudie i grundskolans årskurser 3, 6 och 9. Gbg 1992. Pp. 253.

85. *Valentin González:* On human attitudes. Root metaphors in theoretical conceptions. Gbg 1992. Pp. 238.

86. *Jan-Erik Johansson:* Metodikämnet i förskollärarutbildningen. Bidrag till en traditionsbestämning. Gbg 1992. Pp. 347.

87. *Ann Ahlberg:* Att möta matematiska problem. En belysning av barns lärande. Gbg 1992. Pp. 353.

88. *Ella Danielson:* Omvårdnad och dess psykosociala inslag. Sjuksköterskestuderandes uppfattningar av centrala termer och reaktioner inför en omvårdnadssituation. Gbg 1992. Pp. 301.

89. *Shirley Booth:* Learning to program. A phenomenographic perspective. Gbg 1992. Pp. 308.

90. *Eva Björck-Åkeson:* Samspel mellan små barn med rörelsehinder och talhandikapp och deras föräldrar - en longitudinell studie. Gbg 1992. Pp. 345.

91. *Karin Dahlberg:* Helhetssyn i vården. En uppgift för sjuksköterskeutbildningen. 1992. Pp. 201.

92. *Rigmor Eriksson:* Teaching Language Learning. In-service training for communicative teaching and self directed learning in English as a foreign language. 1993. Pp. 218.

93. *Kjell Härenstam:* Skolboks-islam. Analys av bilden av islam i läroböcker i religionskunskap. Gbg 1993. Pp. 312.

94. *Ingrid Pramling:* Kunnandets grunder. Prövning av en fenomenografisk ansats till att utveckla barns sätt att uppfatta sin omvärld. Gbg 1994. Pp. 236.

95. *Marianne Hansson Scherman:* Att vägra vara sjuk. En longitudinell studie av förhållningssätt till astma/allergi. Gbg 1994. Pp. 236.

96. *Mikael Alexandersson:* Metod och medvetande. Gbg 1994. Pp. 281.

97. *Gun Unenge:* Pappor i föräldrakooperativa daghem. En deskriptiv studie av pappors medverkan. Gbg 1994. Pp. 249, [33].

98. *Björn Sjöström:* Assessing acute postoperative pain. Assessment strategies and quality in relation to clinical experience and professional role. Gbg 1995. Pp. 159.

99. *Maj Arvidsson:* Lärares orsaks- och åtgärdstankar om elever med svårigheter. Gbg 1995. Pp. 212.

(cont.)